ITOGI
Summaries of Scientific Progress

THEORETICAL PROBLEMS IN PHYSICAL AND ECONOMIC GEOGRAPHY
Volume 1

General Editor:

L. S. Abramov

GEOGRAPHY SERIES

Editor-in-Chief

A. A. Nazimovich, Dr. of Geographical Science

G.K. HALL & CO.
Boston, Massachusetts
1974

Copyright © 1974 by G. K. Hall and Co., Boston

All Rights Reserved. No part of this book may be used or reproduced in any manner whatsoever without written permission from the publisher except in the case of brief quotations embodied in critical articles and reviews.

Original Russian text published by the All-Union Institute for Scientific and Technical Information (VINITI) in Moscow in 1973 as a volume of Itogi Nauki. Translation published under contract with the USSR Copyright Agency (VAAP).

Printed in the United States of America.

Library of Congress Catalog Card Number 74-11324
ISBN 0-8161-2008-0

ИТОГИ НАУКИ

ТЕОРЕТИЧЕСКИЕ ВОПРОСЫ ФИЗИЧЕСКОЙ И ЭКОНОМИЧЕСКОЙ ГЕОГРАФИИ

Том 1

Общий редактор:

Л. С. Абрамов

СЕРИЯ ГЕОГРАФИЯ

Главный редактор:

Докт. геогр. наук А. А. Насимович

Г. К. Холл и Ко.

Бостон, США

1974

Авторские права © принадлежат фирму Г. К. Холл и Компания (G. K. Hall and Co.), Бостон, США, 1974 г.

Все права сохранены. Нельзя никаким методом пользоваться никакими частями этой книги или копировать их без написанного разрешения от издательства, с исключением коротких цитат внутри критических статьей и обзоров.

Оригинальный русский текст напечатан Всесоюзным Институтом Научной и Технической Информации в Москве в 1973 как том серии «Итоги Науки». Перевод издан по контракту с ВААП-ом.

Число в каталоге Библиотеки Конгресса (США) 74-11324

Число ISBN 0-8161-2008-0

Папечатано в Соединённых Штатах Америки.

CONTENTS

	Introduction	vii
	Foreword	ix
S. V. Kalesnik	The Subject, System and Classification of the Geographical Sciences	1
A.A. Mintz	The History of Geographical Thought (A Survey of Soviet Literature, 1961–1970)	15
V.S. Preobrazhenskiy	General Physical Geography (Earth Science and Geomorphology) 1967–1970	33
A.A. Mintz	Economic Geography (A Survey of Basic Tendencies, 1966–1970)	79
Yu. G. Lipetz	The Application of Mathematical Methods in Economic Geography	129

INTRODUCTION

This volume is the first in the Itogi series on physical and economic geography. It is one of the several under the general series heading of Geography published by VINITI (All-Union Institute for Scientific and Technical Information) and is devoted to an analysis of publications dealing with the general issues of fundamental theoretical or methodological significance in physical and economic geography, covering the period from 1966 through 1970. Succeeding volumes will bring the reviews of the literature up to date. The initial volume discusses the subject matter, organization and classification of the geographical sciences, the history of geographical thought in the Soviet Union and deals with the direction of development of physical and economic geography as reflected in the literature surveyed. Mathematical methods as applied to the solution of problems in this field are also analyzed. The authors of the sections of Volume 1 are: Acad. S.V. Kalesnik, Dr. A.A. Mintz, Dr. V.S. Preobrazhenskiy and Cand. Yu.G. Lipetz. They are geographers actively engaged in research in the fields they review, and their approach to the subject is evaluated in the Foreword by the Editor, L.S. Abramov.

The Editor-in-Chief of the series is Dr. of Geographic Science A.A. Nasimovich; members of the Editorial Board are: T.V. Goltseya, Dr. D.V. Boriseyvich, O.V. Vitkovsky, S.A. Gabrilova, A.N. Gratsiansky, I.L. Lementyev, N.K. Leparma, Dr. A.D. Lebedyev, K.C. Losyev, Dr. A.A. Mintz, I.I. Parkomenko, Prof. V.V. Pokshishevsky, Ay.G. Pryabtseva, A.A. Tolokonnikova and Dr. V.M. Fridland.

FOREWORD

One of the most significant phenomena of the present time, one which influences many aspects of modern life, is the ever-accelerating rate of scientific and technological progress. It exerts a powerful influence on science itself: it gives rise to changes in content and to an increase in the role of theory. A significant methodological reorganization is occurring in science, due particularly to the introduction of systems approaches and the methods of the exact sciences. Finally, certain tendencies in the development of science are also changing: along with the continuing processes of differentiation, the process of integration, in which separate sciences combine their efforts in order to solve mutual problems, is intensifying. All of these processes are accompanied by an increase in information flow and a precipitous growth in the number of publications, giving rise to the need for creating new means of processing information. Reference publications are one means of dealing with this problem. These very general considerations pertaining to science as a whole take on particular significance when applied to the science of geography.

The scientific and technological revolution is causing the influence of society on nature to increase sharply both in intensity and depth. It is spreading over enormous areas, of the Earth's surface, and ever-increasing amounts of natural resources are being used in industrial production. All of this increases the significance, not only in predicting the course of natural processes subjected to the effects of human productive activity, but also of the possibility of deliberately influencing the direction of these processes, of maintaining the geographical environment and its resources at the proper level, even to the point of constructing a new environment with the parameters required by human society.

Under these conditions the formulation of predictions, to say nothing of the working out of constructive hypotheses, requires more than knowledge of the course of individual processes; what is required is a general theory of the development of nature, its constituents, and the geographical environment as a whole.

An important element of geographical engineering is the allocation of production facilities and population, with their ever-increasing requirements on the geographical environment and the resources of nature. Planning of this sort requires knowledge of the laws of social evolution and the principles governing the allocation of production and its territorial complexes, as well as knowledge of settlement patterns. Also necessary in this regard are methods for objectively appraising natural conditions and resources. Finally, nature and society do not evolve independently of one another, but rather in close mutual interaction, and the depth and breadth of the phenomena embraced in this interaction are becoming greater. The study of the interrelation between society and nature, having as its goal the intelligent control of the processes involved, is taking on ever greater importance in the research efforts of Soviet geographers.

These considerations imply more than simply an increase in the role of the theoretical branches of geography. The problems facing geography are becoming more and more complex, and many of them are being solved through the combined efforts of a number of the separate disciplines of which geographical science is composed. This state of affairs necessitates an intensive reexamination of the geographical sciences with a view towards determining what unites them, towards discovering the integrational processes in the evolution of contemporary Soviet science, and the development of interdisciplinary approaches to planning.

General physical and general economic geography are the spheres of attraction around which other geographical sciences cluster. They are increasingly coming to be regarded not as the two major groupings of the geographical sciences, but as separate geographical disciplines, each with its own objects of investigation. These two basic sciences, which play a guiding and organizing role with regard to the groups of sciences corresponding to them, also differ significantly from the subsidiary physical-geographic and economic-geographic disciplines, and it is hardly necessary to demonstrate that the theoretical branches of these sciences play a particularly important role in the formulation of issues in geographical theory.

These sciences also play an important role in the introduction of new research methods, into the geographical sciences such as systems approaches to the analysis of complex phenomena, modelling and experimentation, and exact methods of studying and fixing observed data.

Naturally, individual branch disciplines apply their own specific methods. The general tendencies with regard to the introduction of these methods, however, are most typically common to all the geographical sciences, and appear precisely in these sciences and in their research and so the experience of the branch disciplines takes on general methodological significance.

All of these factors stimulate a heightened interest in the general geographical disciplines and in the theoretical and methodological questions with which they deal. Yet, until recently, there have been no surveys analyzing the accomplishments of Soviet geography. The reason for this lack is that the branch disciplines, dealing to a greater extent with concrete objects of investigation, have developed more intensively and have yielded a

FOREWORD

greater volume of published results in our country. The general disciplines, on the other hand, have been occupied with research on specific problems, such as physical-geographic and economic-geographic regionalization. Until recently relatively little work has been done on the theoretical aspects of physical and economic geography. Significant changes in this regard occurred during the second half of the 1960s: more attention was devoted to theoretical and methodological questions in science in general, including the general geographical sciences. There was, in addition to the general logic to the development of science, an additional stimulus in this regard in the form of a desire on the part of Soviet geographers to analyze the current theoretical and methodological level of the geographical sciences and to summarize their accomplishments during the first 50 years of Soviet power, especially in the light of Leninist ideas, in celebration of the 100th anniversary of Lenin's birth.

The time has now come to summarize the results of these efforts. For this purpose the ITOGI series, "Summaries of Scientific Progress," in which leading specialists survey the literature of a given field during a given time frame is an entirely appropriate and useful format. Such surveys are more than highly qualified resumes providing orientation in a mass of publications; they permit the reader to summarize the progress in a science, to determine its scientific and methodological level, and to elucidate its relationship to other sciences or the disproportions existing between the various divisions of the same science.

This volume of the ITOGI series is devoted to an analysis of publications dealing with the most general questions in physical and economic geography, those having major theoretical or methodological significance. It covers the question of the subject matter of the geographical sciences, their structure, and their classification; it analyzes publications on the history of geographical thought in the USSR, and surveys the two basic geographical sciences from which the others derive, general physical geography and general economic geography. Of the methods of the exact sciences which have been introduced into geography the most closely examined are those of mathematics, since they are the most universal. The authors of the articles in this collection are major specialists in the field who are actively researching the problems in question: Academician S. V. Kalesnick, President of the Geographical Society of the USSR, V. S. Preobrazhenskiy and A. A. Mintz, Department Heads in the Institute of Geography of the Academy of Sciences of the USSR, and Yu. G. Lipetz, Candidate in Geographical Sciences.

It is to be expected that such competent authors will not confine themselves to a simple survey of the literature, its systematization and classification, but will give, in addition, their evaluation of a particular issue or area of research, and even of science in general. Of course, such evaluation cannot help but be in some degree subjective. Each of the major scientists arrives at his evaluation in his own individual fashion. These articles, therefore, require no special introduction. However, it is clearly necessary to note the differences between them, which do not, nonetheless, destroy the unity of the collection, and to emphasize once again the most general tendencies described in the articles.

Above all, the development of the general branches of geography proceeds unevenly, and the interest shown in the various topics, and, consequently, the number of publications devoted to them, are scarcely the same. This in turn determines a need for frequently surveying individual sciences or topics over whatever period which is most expedient for the corresponding survey. In the present survey therefore, the development of the various sciences is examined over unequal periods of time. Thus, the number of recent publications dealing with the subject matter and methods of the individual sciences has been relatively small, and so the article by S. V. Kalesnik surveys a relatively long time span. A. A. Mintz examines in his article for a 10-year span publications on the history of geographical thought, for a 5-year span basic tendencies in the development of economic geography. Preobrazhenskiy, meanwhile, presents a very concise summary of the extensive literature in general physical geography published only during the last 3 years dealing with the especially intensive development of geomorphology.

The first article in the collection is that of Academician Kalesnik, entitled "The Subject, System and Classification of the Geographical Sciences." This traditional and broad theme continues to be important in geography since, as the author rightly notes, classification fulfills two basic functions. First, it defines the sphere of interest of each science and, thereby, not only clarifies their lines of demarcation, but facilitates their development by fixing their place among the other branches of knowledge and by strengthening their theoretical positions and facilitating the application of practical results. Second, classification unites different but related sciences, defining their areas of interaction and cooperation. As is customary in publications dealing with these questions, the author concentrates his analysis on elucidating the subject matter of each of the geographical disciplines and devotes special attention to questions involving the differentiation of fields of scientific investigation. Integrational tendencies, the combining of many sciences for the purpose of solving problems common to them all, though recently on the increase, have not as yet been discussed extensively in the literature, and this fact, naturally, is reflected in Academician Kalesnik's survey.

In his article entitled "The History of Geographical Thought (A Survey of Soviet Literature, 1961-1970)" Mintz notes that traditional historico-scientific publications continue to predominate in this field and that there has been very little written in the way of general works devoted to the history of geographical thought per se. In most instances contemporary researchers deal with discrete problems and develop the theoretical aspects of their discipline only in passing. Yet a generalized approach would aid in the systemization of work in this area, make it easier for specialists to understand the

FOREWORD

historical development of their fields, and in general further the development of Soviet geography.

The author of the article entitled "General Physical Geography," Preobrazhenskiy, limits his discussion to publications dealing with earth science and geomorphology. He does not discuss here physical–geographic regional geography as it traditionally relates to general physical geography, since he is concerned primarily with methodological rather than with theoretical problems. His survey clearly shows that earth science in the Soviet Union is developing significantly slower than is geomorphology: publications in earth science, aside from a small number of research monographs, are in the nature of textbooks or instructional manuals. This constitutes a significant gap in Soviet science, inasmuch as the achievements of earth science should be accompanied by a corresponding general theoretical development. Geomorphology is undergoing considerably more intensive development, and, as the author notes, is beginning to show internal differentiation.

Mintz's emotionally penned survey of basic tendencies in the development of economic geography shows that, although it is becoming more important, a number of specialists are tending to solve certain practical problems, especially in the area of territorial organization of productive capacities, by means that lie outside the range of economic geography, namely those of the regional divisions of the economic sciences. The author demonstrates the superficial and parochial nature of these tendencies, essentially their harmful effect. At the same time, he expresses a certain disquiet regarding the place and significance of the economic geography sections in the system of the Academy of Sciences and universities.

Lipetz's article "The Application of Mathematical Methods in Economic Geography" deals with the most general and fundamental aspects of prognostics and evaluates the applicability of certain mathematical methods in the solution of problems in economic geography.* The author thus does not confine himself to methodological questions, and in this respect inclusion of his article in a collection devoted to theoretical questions is entirely appropriate.

The content of this volume provides an overall picture of the work Soviet geographers have done in developing the most general theoretical issues in physical and economic geography, as reflected in recent publications. It is hoped that this collection of articles will facilitate the further development of the science of geography.

L. S. Abramov

---

*A special section in the article by Preobrazhenskiy written by T.D. Aleksandrova deals with an analogous problem in the area of physical geography.

THE SUBJECT, SYSTEM AND CLASSIFICATION OF THE GEOGRAPHICAL SCIENCES*

S. V. Kalesnik

Geography is a system of closely interrelated natural and social sciences that emerged during the development and differentiation of an ancient and originally undifferentiated encyclopedic body of knowledge of the physical characteristics, population, and economy of various nations.

Science as a whole, as a body of knowledge about nature, society, and thought, is essentially a unified whole [27, 28], and the development of any of its branches cannot be sharply distinguished from that of related branches. For this reason any classification of sciences must be arbitrary. The interrelated nature of the various branches of knowledge makes it difficult to reduce them to a classification system that could be considered logically perfect; nevertheless it is in practice worthwhile to construct such a system.

In the classification of the geographical sciences the following points must be taken into account.

1. A system of sciences develops historically and arises as a result of differentiation of a broader discipline that was concerned with studying a complex object consisting of a set of simpler objects [42].

2. Geography in ancient times embraced the sum total of existing knowledge concerning the nature of the Earth's surface, as well as the customs and habits of the peoples inhabiting the Earth [12]. Eventually (for the most part during the nineteenth and twentieth centuries) various sections of it divided into independent branches of knowledge, some developing analytically (dealing with individual components of the topography of the Earth's mantle or of separate branches of the national economy—relief, soils, climate, industry, transportation, etc.,) while others preserved a synthetic character, but with a more clearly defined subject

---

*V. V. Pokshishevskiy took part in preparing this article for publication.

matter (natural–territorial or production–territorial complexes, i.e., regional physical and regional economic geography, serve as their subject matter).

3. Along with distinct sections within geography, separate geographical sciences arise as a result of contact with other scientific systems — geology, biology, physics, chemistry, technology, etc., forming divisions of knowledge parallel to them.

4. Certain fields in separating from geography broke away from its structure (ethnography, for example) but most of its daughter sciences did not lose the marks of affinity (i.e., of their common origin), and remained united by a common goal: to study from all sides the geomorphology of the Earth's mantle and features in the distribution and development of human productive activity.

We understand geography, then, not as the encyclopedic science that existed in the past, but as a unified complex of sciences that arose from this once unique body of knowledge. The concrete links uniting human society and its natural environment found actual expression in that both the natural and the social sciences were included in the family of the geographical sciences, although each member of this family is independent of the others, with its own object of investigation.

Man needs for his existence various forms of energy and raw materials, as well as food and water. Sources of energy such as solar radiation, wind, the movement of water in rivers and tides, and components of the environment such as climate, soil, bodies of water, and the plant and animal world, lie within the sphere of the scientific interests of geography. As human society develops its need for different forms of natural resources, its interrelation with the conditions of the natural environment continually increases, as does the importance of geographical investigation of the Earth's surface, the role of geography in the solution of problems of multifaceted and rational exploitation of nature, including its preservation, regeneration, reclamation, and transformation becoming more important [1, 55].

The geographical sciences can be divided into four groups [25]: a) natural geographical or, in the broad sense of the term, physical geographical sciences, to which belong geomorphology, climatology, oceanology (with its subdivision, oceanography), land hydrology, glaciology, geocryology, soil science and geography, biogeography, physical geography in the narrow sense of the term (general earth science); regional physical geography; paleogeography or phenology (the study of the seasonal geomorphic rhythm); b) socio–geographical sciences — the history of geography and of individual geographical disciplines, toponymics and economic geography with all its subdivisions; c) cartography; d) groups of joint geographical disciplines — regional geography, local geography, medical geography, and military geography — which use date of the natural and socio–geographical sciences and cartography, but which also apply for their purposes the methods and results of other branches of knowledge. Parts of the geographical disciplines

THE SUBJECT, SYSTEM AND CLASSIFICATION OF SCIENCES

enumerated above enter into the structure of other sciences (biology, geology, economic, sociology, etc.), since there are no sharp and impermeable boundaries between the sciences.

Although the goals of their investigation are the same, each of the sciences entering into the geographical complex has its own object of investigation, which does not coincide with the objects of the other sciences; each studies its object using the methods required to achieve a deep and comprehensive understanding of it [18].

Each of the geographical sciences has its general theoretical division, its regional division and its applied divisions.* There is no need to emphasize the existence of these fields; they need be isolated only when tradition demands.

Disciplines belonging to secondary education need not be included in the classification.

We will now present, using the above classification, a more extensive catalog of the geographical sciences forming the complex or system of the geographical sciences.

A. THE GROUP OF NATURAL GEOGRAPHICAL SCIENCES

1. Geomorphology [17, 51], in the scope of which are included sections dealing with morphostructures and morphosculptures, and contemporary movements of the earth's crust.

2. Climatology, subdivided into general climatology, the study of the microclimate, climatology of the free atmosphere, climatography, and paleoclimatology.

3. Oceanology, a complex science dealing with the oceans and seas, which developed on the basis of oceanography.

4. Land hydrology, divided into general hydrology, hydrology of rivers, limnology, the study of swamps, hydrogeology, hydrometry, and hydrography.

5. Glaciology, divided into general glaciology, regional glaciology, and paleoglaciology. Glaciology is sometimes subsumed, rightly, under land hydrology. However, in view of the fact that glaciers possess specific characteristics which strongly distinguish them from other objects of hydrological investigation, it is preferable to separate glaciology into a separate subgroup, leaving to land hydrology the study of natural accumulations of liquid water. Land hydrology, oceanology, and glaciology, taken together with hydrogeology, which belongs among the geological sciences, constitute hydrology in the broad sense of the word.

---

*The applied divisions of the geographical sciences are sometimes referred to collectively as applied geography (as a part of the natural geographical sciences they are often termed engineering geography); however, these applied divisions do not constitute independent sciences [55].

6. Geocryology is the science dealing with freezing of the earth's crust, and borders on geography, on the one hand, and soil science, geology, and the engineering sciences, on the other.

7. Soil geography, a part of soil science, developed at the junction of the geographical, biological, and agricultural sciences.

8. Biogeography, divided into botanical geography (bordering botany, which studies the earth's plant cover, or the set of phytocenoses, as a geomorphology) and zoogeography (bordering zoology).

9. Phenology, which forms a part of geography as the science of seasonal geomorphological rhythm.

10. Physical geography [21], divided into general earth science (which studies the general physical–geographic laws of the planet [26, 21]), geomorphology (which studies terrains and their groupings into various taxonomic levels [22, 54]), paleogeography (the study of landscapes from the geological past [36]), historical physical geography (paleogeography of historical times).

B. THE GROUP OF SOCIO–GEOGRAPHIC SCIENCES

11. The history of geography (as a whole and of the individual geographical sciences).*

12. Toponymy (the science of the origins and meanings of geographical names, bordering linguistics and geography).

13. Economic geography [7, 19, 39, 49, 56, 59], divided into population geography [36, 57], the geography of natural resources, industrial geography, agricultural geography, transportation geography, and a new economic–geographical discipline which is now in the process of developing the geography of services [29]. World industrial geography and the geography of territorial–industrial complexes or combined regional economic–geography should be included in this category along with similar multicomponent economic–geography disciplines [31].

C. CARTOGRAPHY

14. Cartography may be divided into cartology, mathematical cartography, cartometry, and map design, drawing, editing, and publishing.

A close bond between geography and cartography has existed since ancient times; for many centuries, in fact, it was considered that the major task of geography consists in map-making. In modern times the

---

*The inclusion of the history of geography (including physical geography) in the socio–geographical sciences is based on the developmental course of every science being conditioned by the development and requirements of society and should be viewed as within the scope of social phenomena. Moreover, geographical discoveries have themselves been important social phenomena [60].

THE SUBJECT, SYSTEM AND CLASSIFICATION OF SCIENCES

functions of geography are incomparably wider. It is nevertheless necessary to emphasize that cartography occupies a special place in the modern family of the geographical sciences, due to cartography being closely related to the mathematical and technical sciences, and is utilized by both the natural and the social disciplines; it deals not only with geographical maps and techniques of mapping, but constitutes a powerful method of comparative investigation and surveying of geographical phenomena both through space and through time, both statically and dynamically [43].

D. THE GROUP OF JOINT DISCIPLINES

It is sometimes convenient to collect in one place facts established by different sciences; in other instances it is necessary to simultaneously illuminate a particular issue from the point of view of a number of branches of knowledge. This leads to the formation of new, joint disciplines of an applied nature that solve their own problems using material from several sciences. In geography the following are disciplines of this sort.

15. Regional geography [6, 13].

16. Local geography.

Regional geography is distinguished from local geography only in that the former deals with the collection and systematization of various physical-geographic, economic–geographical, historical, and other data relating primarily to large territories — continents, countries, and regions, which the latter deals primarily with small territories.

17. Military geography, divided into general military geography (the study of the geographical prerequisites for strategy and tactics) and regional military geography (the characterization of theaters of military operations).

18. Medical geography, studying the influence of the geographical environment on human health (developed at the border of geography and medicine).

Certain branches of knowledge included in this classification of the geographical sciences could also be included in other systems of sciences (hydrology could be included in geology, biogeography in biology, etc.), and this is perfectly acceptable [8, 9, 10, etc.]. If there exists a branch of knowledge whose content lies on the borders of two fields of knowledge, then the true state of affairs can be reflected only by classifying it as part of two systems of sciences. Limiting the position of such a discipline to within the scope of only one scientific sphere would be a procustean operation, contrary to logic and unrealistic.

The existence of interacting analytic and synthetic disciplines is one of the internal conditions for the progress of the entire complex of the geographical sciences. If the synthetic branches of geography are separated from the analytical, then both suffer. In the course of the development of the geographical sciences (and others — geological, biological, etc.), however, the process of differentiating them was more intense, and a balance between the analytic and synthetic parts was not achieved.

The quantitative predominance in the complex of the geographical sciences of the branch, or subsidiary, disciplines and their intensive growth as compared with the slower development of the synthetic members of the complex (easily explained by their greater complexity) caused alarm among some geographers. They began to search for ways and means to strengthen geographical synthesis [41].

In the physical–geographical sciences considerable success was achieved in the area of synthesis in general earth science and geomorphology [14, 20, 22, etc.]. In the economic–geographic sciences regional economic–geography [e.g. 49] became a successful field for generalization. But the boundary between the natural and societal groups of the complex of the geographical sciences proved backward in this respect [16, 33, 34, 44], although the need for interdisciplinary general geographical research based on close interaction between all geographical sciences, including interaction between physical and economic geography, had long been recognized. Indisputable proof of this fact is the development in the system of the geographical sciences of such a branch as regional geography [6].

Under these conditions geographical research intensified through the collective efforts of geographers working in various fields [11, 52, 53, etc.].

For some researchers, however, the major obstacle to the synthesis of geographical knowledge proved to be the individuality of the majority of the disciplines in the complex of the geographical sciences,* each of which has its own object for investigation. In the opinion of these geographers it would be better if geography became a single complex science. Anuchin [5, 3] has taken this position. His thesis – that geography is not only a single complex of sciences, but a single complex science, since all of the geographical sciences have the same general object of investigation (the geographical environment) – has evoked strong objection [2, 37, etc.]. Strangely, geography emerges simultaneously as a plurality (a complex of sciences) and as a unity (a complex or multiple science). The basic arguments against this thesis are the following.

The material world evolves under the action of both general and specific laws. Dialectical materialism deals with the general laws, while other sciences, including geographical sciences, deal with specific laws. But there are both natural and social sciences among the geographical sciences. If it is allowed that they can have a common object of investigation, i.e., that they combine into a single geography, then this object would be such that its specific laws of development should be equally

---

*Because geography is a system, or complex, of sciences, controversy did not arise. A. A. Grigor'ev has proposed that the term "system" be used to donote a set of sciences which study phenomena of the same kind, complex being used to denote a set of sciences studying phenomena of different kinds [10]. It must be agreed that complex is more appropriate for denoting geography as a whole.

specific for nature and for society.* Such hybrids are unknown in Marxist science.

Anuchin [5] attempted, even though provisionally, to identify the geographical environment with the Earth's geomorphological mantle: his intention was that the geographical sciences should in all instances deal with territories in which an interaction between nature and society is occurring. This thesis too was greeted unsympathetically. Konstantinov expressed the basic objection succinctly: "the geographical environment is only that portion of the surface of the earth which is *directly* associated with human life and activity" [32].

Without an artificial fusion of human society with the geographical environment it is impossible to obtain an object of investigation common to all the geographical sciences. Thus, in essence, all geographical monism rests on the assertion that human society enters into the composition of the geographical environment surrounding it. If this assumption is refuted, the entire monistic conception of geography collapses like a house of cards. A society is an association of human beings based on productive relationships. If there are no productive relationships between people, there is no society. Society cannot be divided into two categories, one with and one without productive relations between people, and since productive relations cannot be included in the geographical environment, society as such likewise cannot be included.

Do people themselves, the objects to which human labor is applied, the tools, and results of labor enter into the geographical environments?

In order to demonstrate that this is the case Saushkin, in stating that human beings are a part of nature and consequently a part of the geographical environment, has made the following comparison: the environment of a sprout of rye in a field is not only the soil and the climate, but the other sprouts, and so the field constitutes a plant community [44]. If we think of an individual sprout, then this comparison is valid, but if the entire field of rye is being considered it is not valid, since a field of rye as a single plant community does not constitute its own environment. If we think of an individual human being, other human beings do indeed constitute his environment, but a social, not a geographical, environment. Human beings, although they are a part of nature, do not exist outside productive relations. Consequently, it is impossible for them to break away from society and become an element in the geographical environment.

The tools of labor are the means by which man acts on the geographical environment in order to obtain material wealth (instruments, machines, agricultural equipment, etc.). How can they enter into the composition of

---

*The attempt to fabricate a common object of study for physical and economic geography brought about a strong convergence of Anuchin's theory of geographical monism with the geographical monism of western European and American geographers. The latter is most clearly embodied in these countries in so-called regional geography, which studies differences in the developmental characteristics of nature and society, and so is concerned with the isolation of complex natural–social regions.

the object on which they act? This is no less absurd than the inclusion of human society in the structure of the geographical environment which surrounds it.

The issue with regard to the objects of human labor must be resolved as follows. All natural bodies acted on by man but preserving their typological analogs in virgin nature remain elements of the geographical environment (canals, reservoirs, fields, gardens, pastures, man-made forests, cultivated plants, domestic animals, etc.); all natural bodies changed by human labor to such an extent that they have lost all analogy with any natural elements of the geographical environment are to be considered separate from it. At the same time, objects of labor, tools of production, and products of the material activity of people are elements of society, since they are the product of social labor, existing in a system of material relations and fulfilling specific social functions [32].

Among Anuchin's serious theoretical errors is his assertion that the geographical environment can in certain instances (through production) change in the direction of social development [5]. This remnant of geographical determinism, its accompanying discussion notwithstanding, naturally encountered severe criticism.*

As is evident from the above discussion, the principal innovations proposed by monist geographers have proved invalid. However, it has been shown that geographical synthesis is one of our weak areas. The development of a methodology for such synthesis, in other words, for solving the problems posed in regional geography therefore becomes especially important. Regional geography in its fullest form is the description of the natural characteristics, population, and economy of territories included within national, administrative, and historico-geographic boundaries. The major problem in this field lies in the establishment of the relations existing within such a territory between its natural features, population, and economy and the compilation, on the basis of a scientific elucidation of these relations, of complex geographical characterizations.

Some Soviet geographers propose that since regional geography is not an independent science it should be divided into physical–geographical regional geography and economic–geography regional geography [35, 37]. Their motive in making this proposal is that we deal in this field with two different objects (natural conditions and economic conditions) that cannot, for purposes of scientific research, be combined into one.

In our opinion, the division of regional geography into two branches is unjustified. To do research in regional geography is to synthesize knowledge acquired by the different geographical sciences, i.e., to combine the achieve-

---

*Anuchin evidently agrees with this criticism, since later in 1964 he wrote: "The geographical environment cannot be the cause of the transition from feudalism to capitalism, but through the medium of production it *determines* [emphasis in original] many of the *characteristics* [emphasis added] of feudalism and capitalism" [4]. Such a formulation is incontestably valid, and debate on this issue may be considered settled.

ments of these science but not to fuse them into a single science. A second consideration in this regard is that in the complex of the geographical sciences regional geography is a field destined by the course of development which geography is following to perform a general synthesis of the entire field of geography. For this reason to split up this field, i.e., to introduce a trend of differentiation of sciences into it, would be incorrect. Finally, such division is not necessary from a practical point of view, since it leads to duplication of certain branches of geography (physical geographical regional geography duplicates, in essence, regional physical geography, while economic geographic regional geography duplicates regional economic geography).

Regional geography by itself does not exhaust all possible paths to geographical synthesis,* but, as a traditional form (although requiring further refinement), it may be utilized for these goals with excellent results.

The concept of geography as a system of sciences has recently been subject to revision, and a new conception has been proposed in its place: geography as a science of the developmental laws of geosystems and of the control of them [47]. Geosystems here are understood as an aggregate of independent spacial systems of the natural environment, economy, and population. Since this aggregate includes complexes with different rates of development and different geographic ranges, a geosystem does not have definite boundaries either in time or in space. Geosystems arise on a natural basis, as the result of man's economic activity, and develop according to economic laws. Consequently, the new definition of geography gives rise to two conclusions: 1) the object of geography becomes a nebulous entity devoid of concrete forms (because of geosystem does not have definite boundaries): and 2) geography becomes transformed into economic geography (since a geosystem, as characterized above, is essentially an economic region). Clearly, neither one of these conclusions is acceptable.

* * *

Classification fulfills two basic functions. On the one hand, it delimits sciences and clarifies their spheres of interest, thereby promoting the development of each of them, since a correct understanding of their specific tasks and the place each occupies among other branches of knowledge strengthens the theoretical positions of each science and increases the possibilities for practical application of their achievements. The classification of the geographical sciences formulated above is intended to serve these purposes.

On the other hand, classification unites sciences by outlining the broad area of their interaction, interrelations, and collaboration. Preliminary de-

---

*We should recall the relativeness of the division of the geographical sciences into analytic and synthetic. Branching geographical disciplines are analytic with respect to the characteristic of territorial complexes, since they study not the entire complex but, rather, some component of it (climate, soil transportation, etc.). But in studying their proper objects each of them strives toward synthesis, that is towards clarification of the relations between the different sides of an object.

limitation makes such collaboration especially effective. In the final analysis, the development of the geographical sciences proceeds as a unified, internally interconnected system; the achievements of any individual link in this system inevitably become the property of all other links.

LITERATURE CITED

1. Abramov, L.S. The problem of the interaction of nature and society and the role of the geographical sciences in its solution, Voprosy filosofii, 1965, No. 2.
2. Al'brut, M.I. Errors in the philosophical foundations of "unified geography," Izvestiya (Bulletin) Vsesoyuznogo Geograficheskogo Obshchestva, 1963, No. 5.
3. Anuchin, V.A. Criticism of unified geography (O kritike yedinstva geografii), Moscow, Moscow State University Press, 1961.
4. Anuchin, V.A. Synthesis in geographical science, Voprosy filosofii, 1964, No. 2.
5. Anuchin, V.A. Theoretical problems of geography, Moscow, State Publishing House of Geographical Literature, 1960.
6. Baranskiy, N.N. Area studies and physical and economic geography, Izvestiya (Bulletin) Vsesoyuznogo Geograficheskogo Obshchestva, 1946, No. 1.
7. Baranskiy, N.N. Economic geography. Economic cartography, Moscow, Geografgiz, 1956.
8. Blauberg, I.V. Integrity of the geographical mantle. Philosophical problems of natural science, 3 (The geologic-geographical sciences), Moscow, Moscow State University Press, 1960.
9. Bukanovsky, V.M. Principles and basic features of the classification of contemporary natural science, Perm', 1960.
10. Interaction of the sciences in the study of the Earth, Moscow, Press of the Academy of Sciences of the USSR, 1964.
11. Gerasimov, I.P. Geography in the modern world, Vestnik (Journal) of the Academy of Sciences of the USSR, 1960, No. 12.
12. Gettner, A. Geography, its history, nature, and methods, Moscow-Leningrad, State Publishing House, 1930.
13. Gokhman, V.M. and Ignat'yev, G.M. Area studies. In: Soviet geography, sources and problems (Sovetskaya geografiya. Itagi i zadachi), Moscow, State Publishing House of Geographical Literature, 1960.
14. Grigor'yev, A.A. Some problems of physical geography, Voprosy filosofii, 1951, No. 1.
15. Dzhavakhishvili, A.N. Structure of geographical science, Trudy (Transactions) of Tbilisi University, **58**, 1956.
16. Doskach, A.G., Trusov, Y.P., Fadeyev, E.T. Interaction of nature and society in modern geography, Voprosy filosofii, 1965, No. 4.
17. Yefremov, Yu.K. Role of geomorphology in the geographical sciences, Voprosy geografii, 1950, No. 21.
18. Yefremov, Yu.K. Classification of the geographical sciences. In: Zhizn' Zemli (Life of the Earth), In: Museum of Area Studies, Moscow, Moscow State University, No. 2, 1961.
19. Zhirmunskiy, M.M. Subject matter of economic geography as a science, Izvestiya (Bulletin) of the Academy of Sciences of the USSR, Seriya Geografiya, 1951, No. 3.
20. Zabelin, I.M. Geographical environment, geographical complexes, and the system of the geographical sciences. Izvestiya (Bulletin) Vsesoyuznogo Geograficheskogo Obshchestva, 1952, No. 6.

21. Zabelin, I.M. Basic problems of the theory of physical geography, Moscow, Geografgiz, 1957.
22. Isachenko, A.G. Basic principles of physical-geographical regionalization and the construction of a taxonomic system of terms, Leningrad, Uchenyye zapiski (Scientific Notes) of Leningrad State University, Seriya Geografiya, 1962, No. 317, Issue 8.
23. Kalesnik, S.V. Monism and dualism in Soviet geography, Izvestiya (Bulletin) Vsesoyuznogo Geograficheskogo Obshchestva, 1962, No. 1.
24. Kalesnik, S.V. Results of the renewed debate regarding "unified" geography, Izvestiya (Bulletin) Vsesoyuznogo Geograficheskogo Obshchestva, 1965, No. 3.
25. Kalesnik, S.V. The system of the geographical sciences, Nauchnyye doklady (Scientific Reports) vysshoy shkoly, geol-geograficheskiye nauki, 1959, No. 1.
26. Kalesnik, S.V. General geographical laws of the Earth (Obshcheye geograficheskiye zakonomernosti), Moscow, Mysl' Press, 1970.
27. Kedrov, B.M. Classification of the sciences (Klassifikatsiya nauk), **1**, Moscow, The Publishing House of the Higher Party School of the Academy of Social Studies of the Communist Party's Central Committee, 1959.
28. Kedrov, B.M. Classification of the sciences (Klassifikatsiya nauk), **2**, Moscow, Mysl' Press, 1965.
29. Kovalev, S.A. Geography of consumption and population services, Vestnik (Bulletin) of Moscow State University, Seriya Geografiya, 1956, No. 2.
30. Kolosovskiy, N.N. Scientific problems of geography, Voprosy geografii, 1955, No. 37.
31. Kolotiyevskiy, A.M. Problems in the theory and methodology of economic regionalization as related to the general theory of economic geography (Voprosy teorii i metodiki ekonomicheskogo rayonirovaniya v svyazi s obshchey teoriyey ekonomicheskoy geografii), Zinante Press, Riga, 1967.
32. Konstantinov, F.V. Interaction of nature and society in contemporary geography, Izvestiya (Bulletin) of the Academy of Sciences of the USSR, Seriya Geografiya, 1964, No. 4.
33. Konstantinov, F.V. Nature, society, and contemporary geography (Priroda, obshchestvo, sovremennaya geografiya), Priroda Press, 1964, No. 8.
34. Kotel'nikov, V.L. Influence of society on the geographical environment, Uchenyye Zapiski (Scientific Notes) of the Moscow State Pedagogical Institute, 1958, Issue 3.
35. Markov, K.K. Basic laws of the development of the geographical environment, Vestnik (Bulletin) of Moscow State University, Seriya Geografii, 1950, No. 3.
36. Markov, K.K. Paleogeography (Paleogeografiya), 1st ed., 1951; 2nd ed., revised, Moscow, Geografgiz, Ura, 1960.
37. Pokshishevskiy, V.V. Economic-geographical area studies in the system of the geographical sciences, Izvestiya (Bulletin) of the Academy of Sciences of the USSR, Seriya Geografii, 1960, No. 5.
38. Pokshishevskiy, V.V. Content and basic goals of population geography. In: Population geography in the USSR (Geografiya raseleniya v SSSR), Moscow-Leningrad, 1964.
39. Pokshishevskiy, V.V. Nature of the laws of economic geography, Izvestiya (Bulletin) of the Academy of Sciences of the USSR, Seriya Geografiya, 1962, No. 6.

40. Pomazanov, S.I. Theoretical problems in economic geography and area studies, Izvestiya (Bulletin) of the Academy of Sciences of the USSR, Seriya Geografiya, 1961, No. 3.
41. Ryabchikov, A.M. Interaction among the geographical sciences, Vestnik (Journal) of Moscow State University, Seriya Geografiya, 1964, No. 3.
42. Sadovskiy, V.N. Methodological principles of the investigation of systems, Uchenyye Zapiski (Scientific Notes) of Perm' State University, 1962, No. 41.
43. Salishchev, K.A. Cartographical method of research, Vestnik (Journal) of Moscow State University, Seriya Geografiya, 1955, No. 10.
44. Saushkin, Yu.G. Geographical environment of human society, Geografiya i khozyaystvo, No. 12, 1963.
45. Saushkin, Yu.G. Contemporary system of the geographic sciences in the USSR, Vestnik (Journal) of Moscow State University, Seriya Geografiya, 1959, No. 4.
46. Saushkin, Yu.G. Comparison of networks of the basic economic and tectonic regions of the USSR, Voprosy geografii, 1959, No. 47.
47. Saushkin, Yu.G., Smirnov, A.M. Role of Lenin's ideas in the development of theoretical geography, Vestnik (Journal) of Moscow State University, Seriya Geografiya, 1970, No. 1.
48. Semevskiy, B.N. Problems in the theory of economic geography (Voprosy teorii ekonomicheskoy geografii), Leningrad, Leningrad State University Press, 1964.
49. Semevskiy, B.N. Subject matter, method, and problems of economic geography, Izvestiya (Bulletin) Vsesoyuznogo Geograficheskogo Obshchestva, 1960, Issue 2.
50. Soviet geography in our time (Sovetskaya geografiya v nashi dni), Moscow, Znaniye Press, 1961.
51. Soviet geography during the period of construction of communism (Sovetskaya geografiya v period stroitel'stva kommunizma), Moscow, Geografgiz, 1963.
52. Soviet geography. Results and goals (Sovetskaya geografiya. Itogi i zadachi), Moscow, Geografgiz, 1960.
53. Current problems in geography. Scientific reports by Soviet geographers at the Twentieth International Geographical Congress (Sovremennyye problemy geografii. Nauchnyye soobshcheniya sovetskikh geografov po programme XX Mezhdunarodnogo geograficheskogo kongressa), Moscow, Nauka Press, 1964.
54. Solntsev, N.A. Interrelations between living and dead nature, Vestnik (Journal) of Moscow State University, Seriya Geografii, 1960, No. 6.
55. Sochava, V.B. Practical significance of research and the concept of applied geography, Irkutsk, Institute Report on the Geography of Siberia and the Far East, 1965, Issue 9.
56. Feigin, Ya.G. Subject matter of economic geography, its goals and place in the system of sciences. In: Methodological problems in economic geography (Metodologicheskiye voprosy economicheskoy geografii), Moscow, Ekonomizdat, 1962.
57. Khorev, B.S. Place of population geography and populated points in the system of the geographical sciences, Voprosy geografii, 1952, No. 56.
58. Shchukin, I.S. Place of geomorphology in the system of the natural sciences and its interrelations with complex physical geography, Vestnik (Journal) of Moscow State University, Seriya Geografii, 1960, No. 1.

59. Economic geography in the USSR. History and contemporary development (Ekonomicheskaya geografiya v SSSR. Istoriya i sovremennoye razvitiye), Moscow, Prosveshcheniye Press, 1965.
60. Yugai, R.L. Interaction between the history of geography and historical geography, Izvestiya (Bulletin) of the Academy of Sciences of the USSR, Seriya Geografii, 1970, No. 2.

THE HISTORY OF GEOGRAPHICAL THOUGHT
(A Survey of Soviet Literature, 1961-1970)

A. A. Mintz

The contemporary scientific and technological revolution has given rise to an unprecendented interest in science, not only as a system of knowledge, but as a social phenomenon. Among the many aspects of a scientific creativity that are becoming the object of specialized study by the philosophy of science, the "science of science" [20], the history of science, with its central element, the history of scientific thought, occupies an important place.

In discussing the significance of the history of science in connection with the philosophy of science, Mikulinsky and Rodny have noted that the history of science, as the reflection of socio–historical activity, is a part of general history. At the same time, it is one of the more important means for discovering the laws of knowledge in general, and so is an essential element in the research necessary for further development of materialistic dialectics.

It should be emphasized that historical–scientific research cannot be considered as being of secondary importance, since it is called upon to meet the developmental requirements of the individual sciences, to aid in discovering the most promising ways of solving their current problems and in clarifying and formulating their theoretical bases.

Historico–scientific research is an extremely varied topic. It includes a) investigation devoted to the scientific creativity of individual scientists; b) investigation of the development of a specific scientific problem or some branch of science; c) the history of individual sciences; d) the general history of natural science; and d) historico–scientific analysis of individual elements in science and technology — the evolution of methods of research, the role of theory in the development of a science, etc. [38]. In historico–scientific research it is possible to observe emphasis on either factual characterization (chronicle of events) or elucidation of the laws of development of scientific knowledge, the evolution of ideas, the conflict between different views, the development of concepts and schools of thought, and the abandonment of out-dated theories and the rise of new ones. Inasmuch as this discussion will be confined to the history of scientific thought and will not consider the broader topic of the history of science in general, only works embodying the second approach need be considered.*

The two directions of historico–scientific research described above also gives rise to a specific division of labor among researchers. Works of the first type, which are traditional in the history of science and so, as a rule, are vastly more numerous, are usually written by specialists using for the most part methods characteristic of historical research. Whenever works of this type are written as a digression from their basic topic by representatives of a particular science, the rules of the genre force them also to analyze their material in a way that is historical in terms of its content and methods.

Works specifically devoted to the history of scientific thought are another matter. They are written, as a rule, by major specialists directly involved in the development of important trends of research and in the formulation of theoretical issues. However far back in time the history of scientific thought is traced, it is still viewed through the prism of the theoretical achievements, difficulties, or issues of the present day. Such works, as a rule, show the unmistakable imprint of contemporary ideas, of the theoretical dispositions of the authors, and often continue (and reinforce) contemporary scientific disputes. Analysis of the history of scientific though is sometimes carried out by philosophers who emphasize here the most general laws governing the development of scientific knowledge, i.e., laws passing beyond the bounds of the science in question.

Turning now from these very general considerations to the immediate object of the present survey, i.e., studies relating to the history of geographical thought and published over the past decade,† the following features seem to be the most salient.

In spite of the abundance of historico–scientific works published over the last decade in the genre traditional for the history of geography (description of voyages, discovery of new territories, biographies of scientists, and other forms of personalia, etc.), there were no major and general works (as there were in preceding periods) devoted to the history of geographical thought — foreign or Soviet — and embracing the entire complex of the geographical sciences with their interconnections and interactions.

---

*This division is, of course, arbitrary. On the one hand, in works of the chronicle type, including personalia, there is a considerable amount of information useful for analyzing the evolution of scientific thought as a whole. Works of the second group, in turn, inevitably contain large amounts of factual and descriptive material concerning events in the development of a science, landmarks in the biographies of leading scientists, etc.

---

†The chronological limits of the present survey cannot be rigorously adhered to, since it will sometimes be necessary to refer to works published before 1961, but containing material relevant to the topic under consideration.

At the same time, the undoubted growth of interest in geographic theory that has occurred over the last few years, as a result of the genetic approach to these phenomena inherent in Marxist–Leninist scientific methodology, was accompanied by a more or less profound analysis of the history of ideas and concepts in individual geographical disciplines. It is works of this sort that serve as the basic source material for the present survey.*

The most frequent way in which analysis of the history of geographical thought, the evolution of its most important theoretical concepts, meanings, and ideas occurs, therefore, is that of passing consideration in connection with the solution of contemporary scientific problems rather than special investigations (although, as we will show, there exist several significant exceptions to this rule).

The same may be said, we believe, regarding evaluations made more than ten years ago of the state of research into the history of geographical knowledge. Thus, Lebedev [34] in his widely known *Soviet Geography, Results and Problems* [55], in discussing the numerous Soviet publications on the history of geographical discoveries and investigations, as well as institutions and organizations, noted that "as yet no similiar monographic generalizations have appeared" devoted to analyzing the development of individual geographical sciences, of the history of the development of theoretical conceptions in the spheres of physical–geographic and economic–geographic sciences during specific periods. The situation has not changed fundamentally in the period under consideration here, as Belov's survey, published in 1967 [9], shows.

It should be noted in this regard that in works specifically devoted to the history of geography and intended for teaching on the university level [6, 10, 19, 42, etc.], the traditional aspects of the subject strongly dominate — the chronological exposition of events in the development of the science, description of voyages, interpretation of the activities of leading geographers of the past, etc.

In defining the content of the history of geography contemporary authors, following Anuchin, include not only the nature of the process of accumulation of geographical knowledge, but also the history of the development of theoretical thought.† In practice, however, this very nearly reduces to the description of voyages and biographical sketches.

This tendency is quite evident even in such general material as is presented in the "Istoriya nauki. Personalia (History of Science, Personalia)" section of the abstracts journal *Geografiya (Geography)*. Since this line of

---

*It is unfortunately not possible to cover such works completely, especially those in which the questions of interest to us are touched on only in passing. We are forced to confine our attention to only those works in which the history of geographical thought constitutes the central content, and to the most widely known general theoretical studies.

---

†Nevsky in one such work considers it possible to limit the task of the history of geography to that of studying the development and growth of physical geography, and to reduce geographical knowledge to knowledge concerning the nature of the earth.

analysis is illustrative in nature, we have limited ourselves to a selective coverage of material published during the three years 1966, 1968, and 1970.

There are a total of 360 Soviet publications under this heading for the period in question. We classify the overwhelming majority of them (256) as personalia: various forms of historico–biographical sketches, as well as contemporary anniversary memorials, necrologies, and bibliographical materials. Another major group of publications is devoted to the history of voyages and discoveries (including investigations of ancient monuments, the history of the study of the individual countries and regions, etc.): 81 works fall into this category. Nine were classified as general works on the history of geography, and another nine could in some measure be considered as dealing with the history of geographical thought, ideas, and conceptions. Fourteen works of a personal character, but showing a clear tendency in the direction towards the study of the history of geographical ideas, were arbitrarily placed in a separate category.

Analysis of these works shows that, among those which apparently deal with the history of geography (and so classified under this rubric), only a small number are actually devoted to the history of scientific thought. However, this fact does not permit a final judgment as to the degree to which this question has been studied. Moreover, as will become clear from what follows, the most important materials on the history of geographical thought do not fall under the corresponding bibliographical rubric, which unquestionably makes it difficult to locate them and hinders exhaustive treatment of the subject.

It is clear from the above discussion that materials dealing with the history of geographical thought during the period in question proved to be scattered in publications of various types.

Specialized works in which characterization and analysis of the development of geographical thought constitutes the basic or entire content will be considered first. Such works are few in number, but they are of major importance.

We could find only one short publication which dealt with geography as a whole and which was oriented primarily towards analysis of the development of theoretical conceptions: an article by Zabelin [23]. The appropriateness of classifying it as a specialized work is underscored by the fact that it was published by a leading historico–scientific center: The Institute of the History of Natural Science and Technology of the Academy of Sciences of the USSR.

The article deals with the development of geography throughout the period of its existence, from the rise of spacial concepts in the thought of primitive man up to the present day. However, the title of the article notwithstanding, its author confines his discussion almost exclusively to the evolution of physical geography (as is reflected in the section headings) and divides geography into two independent sciences — physical and economic geography. The methodological problems that such division entails receive notably little attention ("Some Comments Concerning Economic Geography"), although the idea of transforming this science into the study of the technological sphere is advanced.

This article provides support only for the opinion stated above regarding the absence of publications dealing with the history of geographical thought as a whole, not as the sum total of historical sketches of individual disciplines but throughout its entirety. There is reason to believe that the situation is improving, however.

Publications have appeared that deal directly with the history of the development of the two basic geographical sciences – physical geography and economic geography.

With regard to physical geography, the works in question are those of the leading Soviet theoretician of geography, Grigor'ev [16, 17]. A number of the external characteristics of these publications – the chronological structure of the exposition (for prerevolutionary geography, the first half of the 19th and the second half of the 19th and beginning of the twentieth centuries; for Soviet geography, the periods 1917–1928 and 1929–1934, up to the First All-Union Geographical Conference in 1934),* the careful dating and characterization of events (expedition, activities of institutions and individuals, important publications, etc.) – are shared with other works on the history of geographical science. However, Grigor'ev's attention is basically centered on the analysis of the methodology of physical geography into various stages, study of the development of theory (including the theory of separate physical–geographic disciplines), and the elucidation of the most important scientific problems distinguishing one period from another.

Although both works are devoted to Soviet geography, Grigor'ev, especially in his research on the prerevolutionary period, touches on the role of foreign geographical schools. Another characteristic of his research is the attention he devotes to the philosophical aspects of the development of geographical thought, above all to the struggle between materialistic and idealistic world views in geographical theory. This is most clearly evident in his analysis of the methodological reorganization of Soviet physical geography based on a conscious adoption of the dialectical materialist world view at the end of the 1920s and beginning of the 1930s.

A major survey publication, although dealing for the most part on different levels with evolution of scientific thought, was devoted to Soviet economic geography [58]. This work, unlike the one just discussed, was written by a group of authors. The authors making the greatest contribution to the planning and execution of this project were Baranskiy, Nikitin, Pokshishevskiy, and Saushkin, the leading Soviet economic geographers.

This excellent summary consists of several parts and includes major

---

*Unfortunately, Grigor'ev's historico–scientific researches covered only up to 1934 and did not include analysis of more recent stages in the development of physical geography.

survey articles on the history of prerevolutionary [44] and Soviet [53] economic geography, economic cartography [48], and the connections between Soviet and foreign economic geography [45]. A large portion of the book is devoted to biographical sketches, grouped chronologically, of 49 prerevolutionary and Soviet economic geographers (in addition to short descriptions of 66 actively working Soviet economic geographers given in an appendix). Finally, the book contains a bibliography of basic general literature on the history of Soviet economic geography.

In the articles of this book devoted to the history of economic–geographic science, elements of a chronicle of events (chronological grouping of material, information concerning the many events in the history of science, etc.) are combined with methodological analysis, including critical outlines of scientific trends, schools, and the activities of individual scientists. Analysis of the history both of prerevolutionary and, particularly, of Soviet economic geography is characterized by an attempt to display facts in the history of a science against a general historical background — the history of society and social thought in general. Using this approach, the book in question shows quite clearly the relation of economico–geographic thought to the social and economic conditions of each period and, above all, to the state of the object of this science, the national economy, as well as its relation to the general course of scientific thought, and the struggle of ideas in society and in science. The relation between the methodology of Soviet economic geography and the theory and practice of economic planning is especially clearly drawn.

Inasmuch as the article on the history of Soviet economic geography [53] was written by the most active worker in this field and a participant in much sharp methodological debate, it bears the imprint of subjective positions in its evaluations of various events and individuals and, in a number of instances, in its selection of material.

Specialized works in individual geographical disciplines or research trends are few in number. Monographs on the development of the methodology of economic regionalization [12] and the development of geography of cities as a separate branch of knowledge [29] may be mentioned in this regard. In the first instance, however, it is possible to speak only provisionally of specialized historico–scientific research, since the author (Vetrov) gives considerable space to an exposition of his own views on economic regionalization and its place among other disciplines, which are quite different from those of other workers in this field.

A series of articles by Alampiyev is devoted to an analysis of the works of professional revolutionaries as sources of information concerning the development of economico-geographic thought [3, 4, 5].

Nikitin used an original approach in attempting to examine the stages of the development of economic geography through the prism provided by its status at Soviet geographical conferences [43].

Another important group of publications are works of the survey type that summarize the state of geography or its subsidiary disciplines in one aspect or another.

In a number of these works jubilee or anniversary publications occupy an important place and provide a retrospective overview over a greater or lesser period of time. The breadth and depth of the historico–scientific analysis presented in these publications understandably varies depending on the significance of the event being celebrated. In the ten-year period under consideration here the most significant publications of this type were associated with the 50th anniversary of the Great October Socialist Revolution. This celebration was the occasion for an analysis of the many achievements of the Soviet Union during the first half-century of the existence of the Soviet state. The period in question was sufficiently long to include the birth and evolution of the majority of contemporary geographical concepts.

The most basic historico–scientific summing up published during the period of interest to us was the monograph *The Development of the Earth Sciences in the USSR* [50], appearing in the multivolume series *Soviet Science and Technology over the Last 50 Years*, published by the Institute of the History of Natural Science and Technology of the Academy of Sciences of the USSR. This volume was the result of the efforts of a large number of specialists working in various branches of the earth sciences.*

The grouping of the material was organized not according to classification of sciences but according to subject matter ("Earth as a Whole," "Earth's Crust and Surface Mantle," "Land," "The Ocean," "Atmosphere," "Near-Land Space"). Under this system of geographical sciences were treated in the "Land" section. The history of the development of economic geography was touched on only briefly and in conjuction with the study of natural conditions and resources [8]. A general survey of the development of the overall complex of the geographical sciences did not appear. It is true that, unlike the other sections of the book, the "Land" section included an article oriented towards the future and dealing with the further tasks of geography [14], but even here only physical geography was considered.

The most thorough treatment was given to the history of the development during the Soviet period of the individual physical–geographic disciplines: geomorphology [37], climatology [11], glaciology [1], geocryology [26], hydrology [35], soil geography [57], flora [32], and fauna [41]. In each of these outlines a thorough factual coverage (chronicle of events) was accompanied by data on the development of scientific concepts, tendencies, and schools of thought.

Limitations of space on each section and the need to include sufficient-

---

*The concept of earth sciences was used in the sense in which it is used in the organizational structure of the Academy of Sciences of the USSR, i.e., as excluding the biological, social, and engineering sciences. A partial exception was made to this rule in the section entitle "Land," which included outlines of the biological disciplines.

ly broad factual material (dictated by the task of complete and multifaceted coverage of the history of each discipline) severely limited the possibility for deep analysis of the development of scientific thought per se.

Another anniversary publication constituted a major contribution to historico–scientific literature — a collective monograph dedicated to the 125th anniversary of the Geographical Society of the USSR [13]. In spite of the historical sweep of this work, its very conception imposed a serious limitation: only the activities of the Geographical Society were considered. In addition to general information on the Society's activities and the history of its regional investigation, both in our country and abroad, the book includes outlines of the development of geographical and related disciplines [21, *in toto*]. As in the anniversary report discussed above, much space was devoted to detailed documentation packed with factual data of the chronicle of events type (especially in the sections that were structured in a specifically chronological fashion). At the same time, primarily in those sections devoted to the general sciences — physical and economic geography — and written by leading geographers (Isachenko, Wol'f, and Konstantinov), considerable attention was devoted to the development of theoretical conceptions and basic trends, i.e., to the history of geographical thought in the sense of the term used here.

In this group may be included publications of a survey character, oriented toward elucidating the contemporary state of the science, but including, although in different degrees, analysis of the genesis and evolution of contemporary phenomena.

The work covering the most material was a collection of articles published in 1960 by the Geographical Society of the USSR entitled *Soviet Geography. Results and Problems* [51]. In the Foreword the editorial staff described the collection as a scientific reference work having as its goal the comprehensive description of the state of contemporary Soviet geography, its major theoretical achievements, and its unresolved problems.

Many of the leading geographers of our country contributed to this collection presented at the 19th International Geographic Congress (Stockholm, 1960), in which a large number of Soviet scholars took part. At the same time, this book may be considered as a "reply" to the survey by American geographers, *American Geography. Results and Perspectives,* published in 1954 and soon thereafter translated and published in Russia (Moscow, Foreign Literature Publishing House, 1957).*

The content of this collection is extraordinarily broad since the authors succeeded in realizing their intention of covering Soviet geographical science as a whole, both its general and specialized disciplines, as well as major problems, applied research, methodology, and geographical education, and the popularization of knowledge.

As a result of the principle of historicism that dominates Soviet science, most of the articles in the collection contain elements of historico–scientific

---

*"Soviet Geography" was also published in translation in the USA soon after it appeared.

analysis. But lack of space and a general orientation toward characterization of the current state of the disciplines and the problems considered seriously limited the expression of this tendency.

However, the section entitled "The History and Current State of Soviet Geography" and including articles on the synthetic disciplines — the physical geography of land [25], seas and oceans [21], economic geography [30], regional geography [15], cartography [21], and the history of geographical knowledge [34] — is of a somewhat different character. Each of these articles contains a short but informative survey of the development of the branch of geographical knowledge in question, including the development of current theoretical concepts. An analogous approach is used in the articles devoted to the narrower geographical disciplines.

Elements of historico-scientific analysis are encountered in a number of survey works published by VINITI (All-Union Institute for Scientific and Technical Information) in the ITOGI series in Science and Technology, Although works of this type are oriented primarily toward the survey and analysis of current publications (they usually cover only a short period: 1 to 5 years), retrospective analysis frequently goes beyond the framework of the "official" period. This is due in part to the fact that many of the editions of the ITOGI on geography are not part of a series of periodically published surveys, but are rather of a "one-time" character. Thus, the issue devoted to research on problems in economic regionalization [59] was confined to coverage of the period from 1962 to 1964. In a major article by Pokshishevskiy [46], however, considerable attention was devoted to both the historical roots of the Soviet conception of economic regionalization (as opposed to foreign schools of thought) and the fundamental stages in the development of Soviet regionalization. In spite of limited space and the resulting impossibility of detailed chronological coverage, this paper gives a quite thorough characterization of the history of the most important ideas and issues.

In another survey devoted to the development of population geography in the USSR [47], the chronological framework is somewhat broader — from 1961 to 1965. A different format, however — a long bibliography (about 200 entries) and the relatively short survey article by Pokshishevskiy — required that the specifically historical part of the survey be extremely general.* In this case, however, analysis and evaluation of the ways in which scientific thought developed in the geographical discipline under consideration were provided, thus creating a fairly solid genetic base for elucidating the current state of the discipline.

In his survey of physical–geographical regionalization [39] covering the period from 1963 to 1965, Mikhailov based his presentation on the thematic principle. In the introduction to the same edition, however, a short sketch of

---

*This approach is in part to be explained by the fact that not long before publication of this volume in the ITOGI series, survey articles appeared which examined the development of urban geography [25, 31] and rural population geography [40].

the development of research on the problem of physical-geographic regionalization is given. It describes the basic stages through which passed the various research efforts of former periods, current research being examined in the main part of the survey.

The publication of this survey and bibliographic series devoted to industrial-territorial complexes [49] basically covers the period from 1965 to 1969. In the most important articles [56, 18], however, a considerably longer period is covered. Stepanov [56] thoroughly elucidates the entire process of the development of the systems of ideas and methods having to do with the concept of the industrial-territorial complex, beginning with its inception under the direct influence of Leninist theory and practice of the revolutionary transformation of the social and economic structure of the nation. Gusev and Kazansky [18] demonstrate the close connection between theory of industrial-territorial complexes and planning at various stages in the development of the productive capacities of the USSR. This development is examined with in the scope of the general concept of investigation and modeling of territorial systems as a distinct trend in various schools of economic geography.

An extraordinarily large volume of material, though not always adequately developed and concentrated, is contained in the group of papers on the theoretical aspects of the geographical sciences. As a rule, the development of one or another theoretical concept is accompanied by a more or less thorough critical survey of the views of the predecessors in the area in question. Since elements of such an approach are present in any publication of a theoretical character, we may consider only a few examples – monographs that are significant for the general course of development of geographical thought in the USSR and containing important historico-scientific material.*

Among the papers in this group, Anuchin's monograph [7] is of particular interest from the point of view of our subject here. In the first place, it deals with problems having to do with geographical science as a whole; in the second place, it contains a highly elaborated historico-scientific section (the first four chapters of the book are devoted to an exposition of the history of geographical ideas). This section describes, although in a very general way, the development of geographical ideas on a worldwide scale and over an extended historical period – from primitive man until the present time.

Since he regards historico-scientific analysis as the basis for elaborating his own conception of geography (in his terminology, a monistic conception; according to his critics, unified geography), the author devotes special attention to the development of ideas, to the history of geographical thought in the sense of the term used here. Considerable attention is devoted to the connections between geographical methodology and philosophical schools of thought, particularly in regard to the struggle between materialism and

---

*Inasmuch as this approach is associated not with continuous review but with a selection of publications, it inevitably reflects the author's subjectivity.

idealism in philosophy. It must, however, be noted that historico-scientific analysis in Anuchin's book, though of extremely broad scope, is nevertheless of a subordinate nature, oriented towards proof of the validity of the author's monistic concept of geography. It is therefore entirely understandable that the sections devoted to more recent periods (especially the fourth chapter) are often explicitly or implicitly polemical in nature.*

In connection with the exposition of his theoretical conception of physical geography, Zabelin [22] also dealt with the history of geographical thought. In his main chapter ("The Developmental Laws of Geography as a Science") this author emphasizes the difference between his approach to the history of geography and traditional systematized summaries of voyages and discoveries, devoting special attention to the objective laws of the development of science. Both the periodization of the history of geography and the elucidation of the various stages through which it has passed are connected first and foremost with the inception and development of important methodological and philosophical conceptions. Unlike Anuchin, who sees in the history of geographical thought support for the monistic conception of geography, Zabelin concludes from his historical analysis that the division of a once unified geography into two basic sciences — physical geography and economic geography — is an important objective principle in the development of the science. In this instance too, therefore, the presentation of historico-scientific material is in large measure subordinated to the author's theoretical views, the exposition of which constitutes the basic content of the book.

An important historico-scientific section is contained in Isachenko's monograph on theoretical problems in geomorphology [24]. In attempting to show the historical roots of contemporary geomorphology the author touches, although only in a very general way, on the history of geographical concepts, knowledge, and ideas as a whole (a large part of the first chapter being devoted to this subject). In accord with the basic content of the monograph, the prehistory and history of the study of the landscapes, including the stages of development of Soviet geomorphological concepts, receives the most attention.

In the area of economic geography historico-scientific research is closely intertwined with the exposition of the basic theoretical concepts of the science, as was shown above using the example of Saushkin's work [54, 53].

Extensive and varied material on the history of the development of the theory and practical application of Soviet economic regionalization is contained in a two-volume monograph by Alampiyev. The first book [2] is especially worthy of note in this regard; not only is the extensive literature on this subject carefully analyzed, but a considerable quantity of archive material is examined as well. The book contains much factual material with

---

*It should be noted that in the course of the sharp debate evoked by Anuchin's paper the major criticism was directed towards the positive part of Anuchin's concept. The historico-scientific sections of his book were not rejected, especially their factual parts.

regard to the theory, methods, and patterns of economic regionalization, subject to the critical evaluation and analysis of the author, one of the leading Soviet theoreticians in this area of economic geography.

Essentially the same issues in the area of economic regionalization were formulated in a work by Saushkin [58], which, however, differs considerably from Alampiyev's monograph with regard to the quantity of material presented and the format of the presentation, since it was written for delivery as a series of lectures.

It should be noted that both of these works are quite explicitly bear the imprint of the sharp debate on methodological issues in economic regionalization and economic geography as a whole, in which their authors took part.*

Elements of the historico-scientific approach to the problems of economic regionalization with regard to the development of theory in this field may be found in a posthumously published and incomplete book by Kolosovsky [28] and in a monograph by Kolotiyevsky [27].

In the largest category of publications — personalia — important elements in the history of geographical thought are likewise encountered rarely, especially if they are analyzed in large batches. They are, of course, more significant the greater is the contribution of the geographer in question, although the depth of the author's analysis also plays an important role.

It should be noted that several publications, especially those associated with anniversary celebrations in honor of leading scientists, held not only in the Soviet Union but also by international organizations (UNESCO and others), have recently included serious analysis of the contribution made by the man in question, of the significance of his ideas for his epoch and for ours. The many publications in honor of the 200th birthday of Alexander Humboldt, for example, were of considerable significance in geography. Discussion and evaluation of these works, however, is not possible within the limits of this survey.†

Thus, to summarize this survey of literature dealing with the history of geographical thought, it must be noted once again that there have been very few publications devoted to the development of geography as a system of sciences, or of its two basic divisions — physical and economic geography. Future works of this sort should facilitate planning and prediction both in geography as a whole and in its component branches, as well as increase the interaction between them, which is indispensable for the fulfillment of the tasks facing Soviet geography.

---

*In particular it should be noted that, in Saushkin's book, a special section is devoted to criticism of Alampiyev's book, which appeared somewhat earlier.

---

†An interesting example of an anniversary publication with a high information content is Konstantinov's paper written in honor of the 100th birthday of Semonov-Tyan-Shansky [32].

LITERATURE CITED

1. Avsyuk, G.A. Study of glaciers and snow cover and the development of glaciology. In: The development of the Earth sciences in the USSR (Razvitiye nauk o zemle v SSSR), Moscow, Nauka Press, 1967, pp. 367-378.
2. Alampiyev, P.M. Economic regionalization of the USSR (Ekonomicheskoye rayonirovaniye SSSR), Moscow, Gosplanizdat, 1959.
3. Alampiyev, P.M. From the history of Marxist-Leninist economico-geographical thought, Izvestiya (Bulletin) of the Academy of Sciences of the USSR, Seriya Geografiya, 1968, No. 2, 99-110.
4. Alampiyev, P.M. From the history of Marxist-Leninist economico-geographical thought (on regional works of Marxists in exile), Izvestiya (Bulletin) of the Academy of Sciences of the USSR, Seriya Geografiya, 1968, No. 5, 103-118.
5. Alampiyev, P.M. From the history of Marxist-Leninist economico-geographical thought (field work of exiled Marxists), Izvestiya (Bulletin) of the Academy of Sciences of the USSR, Seriya Geografiya, 1969, No. 5, pp. 118-127.
6. Antoshko, Ya.F. History of the geographical study of the Earth. Development of geographical knowledge in the nineteenth and early twentieth centuries (Istoriya geograficheskogo izucheniya zemli. Razvitiye Geograficheskikh znaniy v. XIX– nachale XX vekov), Lecture texts, Moscow, Moscow State University, 1968.
7. Anuchin, V.A. Theoretical problems of geography (Teoreticheskiye problemy geografii), Moscow, Geografgiz, 1960.
8. Armand, D.L., Mintz, A.A., Preobrazhenskiy, V.S., Richter, G.D. Complex study of natural conditions and resources and the development of physical and economic geography. In: Development of the Earth sciences in the USSR (Razvitiye nauk o zemle v SSSR), Moscow, Nauka Press, 1967, pp. 291-317.
9. Belov, M.I. Soviet historico-geographic research. Some results and perspectives, Izvestiya (Bulletin) Vsesoyuznogo Geograficheskogo Obshchestva, **99,** 1967, No. 5, 398.
10. Belozerov, S.T. Development of geography in Russia. Synopsis of lectures on the history of geography (Razvitiye geografii v Rossii. Konspekt lektsii po istorii geografii), Odessa, Odessa State University, 1965.
11. Budyko, M.I., Drozdov, O.A. Basic directions in the development of climatology. In: Development of the Earth sciences in the USSR (Razvitiye nauk o zemle v SSSR), Moscow, Nauka Press, 1967, pp. 354-366.
12. Vetrov, A.S. Basic stages in the development of the methodology of economic regionalization. Economic regionalization as a method and as a science (Osnovnyye etapy razvitiya metodologii ekonomicheskogo rayonirovaniya. Ekonomicheskoye rayonirovaniye kak metod i kak nauka), Chelyabinsk, Yuzhno-Ural'skiy Press, 1967.
13. The Geographical Society over the last 125 years (Geograficheskoye obshchestvo za 125 let), Leningrad, Nauka Press, 1970.
14. Gerasimov, I.P. Further tasks of geography in the system of the Earth sciences and immediate perspectives on the development of new lines of scientific research. In: Development of the Earth sciences in the USSR (Razvitiye nauk o zemle v SSSR), Moscow, Nauka Press, 1967, pp. 441-452.

15. Gokhman, V.M., Ignat'yev, G.M. Area studies. In: Soviet geography. Results and goals (Sovetskaya geografiya. Itogi i zadachi), Moscow, Geografgiz, 1960, pp. 64-72.
16. Grigor'yev, A.A. Development of physical-geographic thought in Russia (19th and early 20th centuries). A short outline (Razvitiye fiziko-geograficheskoy mysli v Rossii (XIX-nachalo XX v.) Kratkii Ocherk), Moscow, Press of the Academy of Sciences of the USSR, 1961.
17. Grigor'yev, A.A. Development of theoretical problems of Soviet physical geography 1917-1934 (Razvitiye teoreticheskikh problem sovetskoy fizicheskoi geografii 1917-1934), Moscow, Nauka Press, 1965.
18. Guseva, V.D., Kazanskiy, N.N. Territorial economic planning and complexes of productive resources. In: Results in science (Itogi Nauki), No. 8, Productive-territorial complexes, Moscow, All-Union Institute of Scientific and Technical Information, 1970, pp. 45-52.
19. Dement'yev, V.A., Andryushchenko, O.N. History of geography. Part I. Geography in ancient and medieval times (Istoriya geografii. Ch. 1. Geografiya v drevniye i sredniye veka), Minsk, Press of the Ministry of Higher and Intermediate Specialized and Professional Education in the BSSR, 1962.
20. Dobrov, G.M. Science of science. An introduction to a general philosophy of science (Nauka o nauke. Vvedeniye v obshcheye naukoznaniye), Kiev, Naukova Dumka Press, 1966.
21. Dobrovol'skiy, A.D. Physical geography of the seas and oceans. In: Soviet geography. Results and goals (Sovetskaya geografiya: Itogi i zadachi), Moscow, State Geographical Literature Press, 1960, pp. 38-48.
22. Zabelin, I.M. Theory of physical geography (Teoriya fizicheskoy geofrafii), Moscow, Geografgiz, 1959.
23. Zabelin, I.M. Evolution of geographical science. In: Outline of the history and theory of science (Ocherki istorii i teorii razvitiya nauki), 1969, pp. 325-347.
24. Isachenko, A.G. Fundamentals of geomorphology and physical-geographic regionalization. A textbook for university students (Osnovy landshaftovedeniya i fiziko-geograficheskoye rayonirovaniye. Uchebnoe posobiye dlya studentov universitetov), Moscow, Vysshaya Shkola Press, 1965.
25. Kalesnik, S.V. Physical geography of dry land. In: Soviet geography. Results and goals (Sovetskaya geografiya. Itogi i zadachi), Moscow, Geografgiz, 1960, pp. 27-37.
26. Kachurin, S.P. Geocryology. In: Development of the Earth sciences in the USSR (Razvitiye nauk o zemle v SSSR), 1967, pp. 379-386.
27. Kolotiyevskiy, A.M. Problems in the theory and methodology of economic regionalization in the context of the general theory of economic geography (Voprosy teorii i metodiki ekonomicheskogo rayonirovaniya v svyazi s obshchey teoriey ekonomicheskoy geografii), Riga, Zinatne Press, 1967.
28. Kolosovskiy, N.N. Theory of economic regionalization (Teoriya ekonomicheskogo rayonirovaniya), Moscow, Mysl' Press, 1969.
29. Konstantinov, O.A. History of the geography of cities in the USSR as a separate branch of geographical knowledge. In: Materials on population geography (Materialy po geografii naseleniya), Leningrad, Geographical Society of the USSR, Economic Geography Section, 1962.

30. Konstantinov, O.A. Economic geography. In: Economic geography. Results and goals (Sovetskaya geografiya. Itogi i zadachi), Moscow, Geografgiz, 1960, pp. 49-53.
31. Konstantinov, O.A. Geographical study of urban populations in the USSR. In: Population geography in the USSR. Basic problems (Geografiya naseleniya v SSSR. Osnovnyye problemy), Moscow-Leningrad, Nauka Press, 1964.
32. Konstantinov, O.A. In honor of Veniamin Petrovich Semenov-Tyan-Shanskiy's Hundredth Anniversary, Izvestiya (Bulletin) Vsesoyuznogo Geograficheskogo Obshchestva, 1970, No. 5, pp. 473-483.
33. Lavrenko, Ye.M. Study of the plant cover and the development of geobotany. In: Development of the Earth sciences in the USSR (Razvitiye nauk o zemle v SSSR), Moscow, Nauka Press, 1967, pp. 419-430.
34. Lebedev, D.M. Research on the history of geographical knowledge. In: Soviet geography. Results and goals (Sovetskaya geografiya. Itogi i zadachi), Moscow, Geografgiz, 1960, pp. 89-101.
35. Lvovich, M.I. Development of the geographical approach in the study of bodies of fresh water. In: Development of the Earth sciences in the USSR (Razvitiye nauk o zemle v SSSR), Moscow, Nauka Press, 1967, pp. 387-400.
36. Markov, K.K. General physical geography. Theory, exact methods of research, and applications in the national economy. In: Soviet geography during the period of the construction of communism (Sovetskaya geografiya v period stroitel'stva kommunizma), Moscow, Geografiya Press, 1963, pp. 27-37.
37. Meshcheryakov, Yu.A. Study of the topography of the Earth's surface and the development of geomorphology. In: Development of the Earth sciences in the USSR (Razvitiye nauk o zemle v SSSR), Moscow, Nauka Press, 1967, pp. 338-353.
38. Mikulinskiy, S.R., Rodnyy, N.I. History of science and the philosophy of science. In: Outline of the history and theory of the development of science (Ocherki istorii i teorii razvitiya nauki), Moscow, Nauka Press, 1969, pp. 35-36.
39. Mikhaylov, N.I. Fiziko-geograficheskoye rayonirovaniye. In: Results of science. Geography of the USSR (Itogi nauki. Geografiya SSSR), Issue 4, Moscow, All-Union Institute of Scientific and Technical Information, 1967.
40. Mintz, A.A. Current state of research on the geography of the rural population of the USSR. In: Population geography in the USSR. Basic problems (Geografiya naseleniya v SSSR. Osnovnyye problemy), Moscow-Leningrad, Nauka Press, 1964.
41. Nasimovich, A.A. Zoogeography. In: Development of the Earth sciences in the USSR (Razvitiye nauki o zemle v SSSR), Moscow, Nauka Press, 1967, pp. 431-440.
42. Nevskiy, V.V. Manual for a course in the history of geography (Metodicheskoye posobiye po kursu istorii geografii), Leningrad, Leningrad State University Press, 1966.
43. Nikitin, N.P. Economic geography at conferences of Soviet geographers, Ucheniye zapiski (Scientific Notes) of the Moscow State Lenin Pedagogical Institute, Problems of physical and economic geography (Voprosy fizicheskoy ekonomicheskoy geografii), Moscow, 1964.
44. Nikitin, N.P. Prerevolutionary economic geography. In: Economic geography in the USSR (Ekonomicheskaya geografiya v SSSR), Moscow, Prosveshcheniye Press, 1965, pp. 9-53.

45. Pokshishevskiy, V.V. Relations and contacts between Russian prerevolutionary and Soviet economic geography and foreign economic geography. In: Economic geography in the USSR (Ekonomicheskaya geografiya v SSSR), Moscow, Prosveshcheniye Press, 1965, pp. 215-240.
46. Pokshishevskiy, V.V. Economic regionalization in the USSR. Survey of Soviet research on problems of economic regionalization from 1962 to 1964. In: Results in science. Geography of the USSR (Itogi nauki. Geografiya SSSR), Economic regionalization in the USSR (Ekonomicheskoye rayonirovaniye SSSR), Issue 2, Moscow, All-Union Institute of Scientific and Technical Information, 1965, pp. 7-45.
47. Pokshishevskiy, V.V. Population geography in the USSR (Geografiya naseleniya v SSSR), Results in Science. Geography of the USSR (Itogi nauki. Geografiya SSSR), Issue 3, Moscow, All-Union Institute of Scientific and Technical Information, 1966.
48. Probrazhenskiy, A.I. Economic cartography. In: Economic geography in the USSR (Ekonomicheskaya geografiya v SSSR), Moscow, Prosveshcheniye Press, 1965, pp. 194-214.
49. Productive-territorial complexes (Proizvodstvenno-territorial'nyye kompleksy. Results in science. Geography of the USSR (Itogi nauki. Geografiya SSSR), Issue 8, Moscow, All-Union Institute of Scientific and Technical Information, 1970.
50. Development of the Earth sciences in the USSR. Fifty years of Soviet science and technology (Razvitiye nauk o zemle v SSSR. Sovetskaya nauka i tekhnika za 50 let, 1917-1967), Moscow, Nauka Press, 1967.
51. Salishchev, K.A. Cartography. In: Soviet geography. Results and goals (Sovetskaya geografiya. Itogi i zadachi), Moscow, Geografgiz, 1960, pp. 63-78.
52. Saushkin, Yu.G. Lectures on the economic regionalization of the USSR (Lektsii po ekonomicheskomu rayonirovaniyu SSSR), Moscow, Moscow State University, Mimeograph, 1960.
53. Saushkin, Yu.G. Soviet economic geography. In: Economic geography in the USSR (Ekonomicheskaya geografiya v SSSR), Moscow, Prosveshcheniye Press, 1965, pp. 54-193.
54. Saushkin, Yu.G. Introduction to economic geography. A textbook for university students (Vvedeniye v ekonomicheskuyu geografiyu. Uchebnoe posobiye dlya studentov universitetov), Moscow, Moscow State University Press, 1970.
55. Soviet geography. Results and goals (Sovetskaya geografiya. Itogi i zadachi), Moscow, Geografgiz, 1960.
56. Stepanov, M.N. Socialist territorial organization of productive resources and the development of concepts on productive-territorial complexes. In: Results in science. Geography of the USSR, Issue 8. Productive-territorial complexes (Itogi nauki. Geografiya SSSR. Proizvodstvennyye-territorial'nyye kompleksy), Moscow, All-Union Institute of Scientific and Technical Information, 1970, pp. 16-44.
57. Fridland, V.M. Study of the soil cover and the development of soil science. In: Development of the Earth sciences in the USSR (Razvitiye nauki o zemle v SSSR), Moscow, Nauka Press, 1967, pp. 401-418.

58. Economic geography in the USSR. History and current development (Ekonomicheskaya geografiya v SSSR. Istoriya i sovremennoye razvitiye), Moscow, Prosveshcheniye Press, 1965.
59. Economic regionalization of the USSR. Results in science. Geography of the USSR (Economicheskoye rayonirovaniye SSSR. Itogi nauki. Geografiya SSSR), Issue 2, Moscow, All-Union Institute of Scientific and Technical Information, 1965.

GENERAL PHYSICAL GEOGRAPHY
(Earth Science and Geomorphology)
1967-1970

V. S. Preobrazhenskiy

In considering developments in the fields of physical geography (earth science & geomorphology) during the period from 1967 to 1970 it must be kept in mind that they proceeded as part of the scientific and technological revolution of the transformation of science into a directly productive force. The scientific and technological revolution has sharpened a number of previously formulated problems and given rise to new ones. It has underscored with particular force not only the acquisition of new resources but also the economic control, control of the natural environment of human life, and the control of natural resources. For this reason increased attention has been devoted to theoretical and methodological research, which plays an important role in the solution of applied and regional problems. This research has given rise to the rapid development of systems-structural and information approaches, without which the prediction of changes in and control of the natural environment are practically impossible.

These years saw a rapid increase in the number of links between our science and other sciences and also with praxis. Contacts widened primarily with construction enterprises and public health; contacts with agriculture also increased. Interaction with other primary geographical sciences, economic geography, medical geography, and population geography intensified.

As a result of this trend not only the above mentioned theoretical conceptions developed, but there also appeared new conceptions (for example, natural complexes as information systems; the unity of the continuity and discreteness of geosystems) and new approaches (the systems-structural, for example); mathematical, physical, and chemical methods were widely introduced, and methods for studying the natural environment of the earth from outer space were developed.

The transformation of science into a directly productive force stimulated the development of approaches based on the scientific study of science itself, the search for ways of controlling both the subject matter and the quality of physical-geographic research.

THE HISTORY AND CURRENT STATE OF GENERAL PHYSICAL GEOGRAPHY

Many authors, publishing both individually and collectively, devoted attention to these topics during the period from 1967 to 1970. The following questions received particularly close attention.

– the role of Leninist ideas in the development of physical geography, the formation of its constructive orientation, the overcoming of crisis situations [92], manifestations of idealism, neopositivism, and reductionism, i.e., the reduction of complex movements to simple ones [125], the development of research on natural phenomena [249, 31], the organization of geographical services [209], the development of conservation concepts [5]

– progress in the development of theoretical conceptions, methodologies, territorial research, and systems of scientific organizations during the period of Soviet power [50, 23, 126, 50, 51, 310, 56]

– analysis of changes in physical geography associated with the scientific and technological revolution and giving rise to a broadening of the subject matter of its research in the direction of man–nature problems [63]; evaluation of the current period as a transition stage, as reflected in the increase in the complexity of subject matter, in the broadening of the bond between science and praxis, in the intensification of the interaction between sciences, in the development and deepening of theoretical investigations, in the vigorous development of systems methods [204, 19, 23, 52, 105, 106]

– analysis of changes occurring in physical geography during particular stages of the current period [257, 3, 177, 205, 120, 54, 77, 245, 90, 121]

– the growth of complex physical-geographic and topographical research in the republics of the Soviet Union (for example [107, 27, 83, 179, 157, 261])

– the history of the study of zonality [159].

As in the past, much attention was devoted to territorial discoveries [50, etc.]. At the same time the concept of considering as geographical discoveries not only the disclosure of various elements of the geographical mantle, but also the elucidation of the objective laws governing its structure and development [263] established itself. In this connection ideas concerning the subject matter of the history of physical geography were clarified: the opinion was expressed that physical geography is a system consisting of two disciplines, the history of discoveries and the study of the earth and the development of the theory of physical geography [100]. As the above summary shows, however, research methodology and the organization of physical geography as a science and of physical-geographic facilities have already been drawn into the sphere of the history of physical geography.

Analysis of the literature indicates that the development of the two components of general physical geography proceeded unevenly during this period: it was more intensive in geomorphology, to which many more publications were devoted than to earth science.

EARTH SCIENCE

For earth science — the study of the geographical mantle as a whole — these years were a period of broadening rather than deepening. The basic types of publication continued to be the textbook [169, 184, 64, 127] and the journal article. New tendencies were also observable — the appearance of specialized collections of articles (for example, *Problems in Planetary Geography*, Moscow, Moscow State University Press, 1969) and individual monographs [173, 235].

The introduction into the individual geographical sciences of the previously formulated (but only recently republished) concepts of Verniadskiy (*The Biosphere*, Moscow, Mysl' Press, 1968) and Grigor'yev [77, 79] continued; these ideas concerned the integrity of the geographical mantle or biosphere. Workers in these sciences are showing an increased interest in analyzing the unity of the processes that occur in the geographical mantle. These tendencies have developed markedly not only in the biological branches of the geographical sciences (geobotany, zoogeography, and soil science) where they were already strong, but also in its geophysical branches: hydrology [127, 13], geomorphology (the study of morphostructures and morphosculptures carried out by Gerasimov and his students), and glaciology [134]. In addition, the ecological branches of biogeography continued to develop rapidly.

Research on the interaction of the components of the geosphere, important for the elaboration of a unified theory of the processes occurring in the geographical mantle [251], broadened in scope on the strength of physical and mathematical analysis of the ocean-atmosphere-land [279] and atmosphere-undersurface-galciation interactions [237].

The study of global anthropogenic factors, especially thermal pollution [99, 236] and chemical substances [71] in the atmosphere intensified.

The basic concept of earth science — the integrity or unity of the geographical mantle as a complex dynamic system — continued to emerge as one which unites the individual branch sciences into a common family of geographical sciences (see, for example, [14, 252]). The individual sciences made an important contribution during this period to the data and empirical laws established by earth science. It is sufficient to note the global surveys of the morphostructures and morphosculptures of the Earth [221], the circulation of natural waters, data from studying bioproductivity in connection with energy resources and heat–moisture relationships, and the investigation of glaciation problems in connection with atmospheric circulation and preglacial topography.

The concept of the integrity of the geographical mantle emerged as a theoretical premise in the design and evaluation of projects involving the transformation of nature [60] and as the fundamental postulate of scientific conservation schemes [5].

All of the above supports Kalesnik's conclusion [124] which, in evaluating the current state of research, asserts that there is no longer any need to justify the existence of general earth science (page 4).

Several new tendencies can be observed in the development of earth science. Among them the attempt to transcend geocentrism seems quite natural for geography. More and more researchers are turning their attention to the relation of the Earth and its geographical mantle to the cosmos [99], to solar rhythms [224]. This tendency was accompanied by a broadening and sharpening of the concept of geographic space, an attempt to find common elements in the geographical mantle of the earth and similar formations on other planets [106]. This group of non- or trans-geocentric ideas may be associated in some measure with the attempt to consider the geographical mantle not only as a system, i.e., as a complex and heterogenous whole, but as a simple physical body — an element of a more complex system [239].

Of fundamental importance for earth science were attempts to put the unity of land geography and oceanology on a new basis [158, 160, 226]. The geographical mantle is a unified whole, but geographers have often limited their investigations of it to the treatment of land. The maps of the world published in most atlases — maps of geological formations and flora and fauna — are mostly only maps of a small part of the geographical mantle, i.e., of land, hence the gaps and errors that occur in concepts of the spacial laws governing many phenomena.

An important step forward was the first summary of the principles established by earth science and the individual physical–geographic sciences up to the present time [124].

Attempts to establish earth science on a paleogeographical basis [190, 37] were important and productive, and much needed. An interesting hypothesis was advanced concerning geomorphogenesis as an important stage in the evolution of the earth [156]. A logical development of ideas previously published by Richter regarding the need to devote more attention to winter processes was proposal to consider winter science as a branch of knowledge having to do with the structure and development of the geographical mantle during the winter [194].

Great progress was made in the division of earth science bordering geomorphology and known as the *study of the differentiation of the geographi-*

cal mantle; the concept of differentiation is usually presented as the theoretical basis of regionalization. Particularly noteworthy here is the transcendence of the traditional issues in this area, such as zonality, tectonics, the relation between heat and moisture, etc. Among the problems discussed over the last few years the following are worthy of special note.

— the relation between phenomena of continuity and discreteness in the geographical mantle [183, 24, 19, 21]

— nonzonal elements in the geographical mantle as discussed in the development of concepts regarding symmetry and dissymmetry [158, 277, 278] and of concepts regarding the role of tectonic movements in the formation of the most important properties of major natural complexes

— similarities and differences in the laws governing differentiation of land and oceans [55, 158, 161, 169, 226]

— the role of the time factor: attempts to establish sequences of natural complexes [247]

— regularities in the formation of contemporary structures in the geographical mantle and the evolution of its zonal morphology, studied using inferences deriving from paleogeographical research [37].

At the same time the following research directions were energetically pursued.

— deepening of the analysis of zonal differentiation processes; the work of Grigor'yev and Budyko, involving the accumulation of new data concerning the interaction between heat and moisture ratios and the bioproductivity of zonal landscapes, continued [227, 231]; work on geochemical problems in zonality [70, 198]; a number of general geomorphological zones were elaborated (Lukasheva and Ryabchikov, diagram in *The World Atlas of Physical Geography,* 1964 [173, 250]; the attempt to use satellite data in the analysis of zonal strucutre [44]

— the spread of the zonal-belt approach to the analysis of coastal areas [154] and oceans [55]

— the accumulation of data on the occurrence of vertical zonality in various mountain systems [30, 255, 139, 12, Dobrovolskiy, Rzhansinskaya, 1967] and investigation of its relation to zonality [269]

— deepening of the analysis of morphology of natural complexes [175, 139, 225]; see also the section "The Dynamics and Development of Landscapes" in the present article, in which more attention is devoted to declivities [254, 46, 282] and the influence of contiguous complexes [283, 235].

To sum up, earth science as a whole developed during the period in question in a single trend; research in individual earth sciences, however, was a significant contribution to this development. Internal differentiation within earth science was slight (with the exception of the study of the differentiation of the geographical mantle, which began to develop as a distinct field) and much less pronounced than in geomorphology. It is possible that this was caused to some degree by the relative lack of theoretical research devoted to individual unsolved problems in earth science. Among these problems can be found, in particular, the creation of a numerical model for the entire geographical mantle, the energetic, metabolic, and informational relations between its components on the one hand, and the major territorial divisions (continents, oceans) on the other.

GEOMORPHOLOGY

Geomorphology is the science of natural territorial complexes or features and it developed with considerable intensity during this period. The subject matter of publications in this field broadened significantly, many of them in the form of thematic collections of articles [57, 256, 58, 59, 152] and journal articles. A bibliography of Soviet literature has been published [241].

During this period the links between geomorphology and certain branches of agriculture (construction and public health), as well as scientific disciplines in the economic, engineering, and medical–biological (medical geography) fields, strengthened and actively broadened. Considerable work was done on nature regionalization of specific areas. The concepts dealing with the morphological structure of landscapes previously formulated by Solntsev and his students continued to be adopted. The number of research projects dealing with specific regions increased, accompanied by publication of small-scale geomorphic maps in complex atlases and of textual geomorphic descriptions. Stationary research facilities expanded and new discoveries in the physics and chemistry of landscapes were made. Certain questions in the classification of natural complexes were clarified. Observation and fixation methods were improved.*

New tendencies also appeared in publications of the last few years. Geomorphology was enriched by the concept of the *natural complex as a self-regulating information system.* The concept of a dialectical association between continuity and discreteness in the structure of natural complexes was advanced. A dynamic structural approach to elementary geosystems developed, and attempts to apply this concept to the analysis of such systems were made. The possibilities for applying methods of mathematics and logic and the principles and methods of modeling to the study both of natural complexes themselves and of the processes by which they are perceived and changed were widely studied.

As a result of this broadening of the sphere of topics studied, it became

---

*Similar processes can be observed in the development of foreign geomorphology traditionally strong in German-speaking countries (see [94]). The introduction of geomorphic approaches in French physical geography characteristerically has been through theoretical channels [288, 306] and in England primarily through applied research [297, 292]. In order to note these similarities we will refer in several places to foreign publications, with no attempt to fully represent the development or the current state of foreign geomorphology.

possible to define contemporary geomorphology as the science dealing with original and transformed natural complexes of all orders. It seeks to elucidate the principles governing the formation, dynamics, development, self-regulation, and distribution of natural complexes, their dynamic and morphological structure, their differentiation and integration, the internal relations between their components that give rise to their integrity, the exchange of substances, energy, and information between landscapes of equal and different orders, and their classification; it studies the laws governing the interaction between natural complexes and engineering structures, agrotechnical procedures and other forms of human influence; it seeks to predict spontaneous and anthropogenic changes in landscapes, and to develop a system of methods for studying and controlling landscapes of various territories.

Together with other sciences geomorphology provides technological (agronomical, engineering, military, etc.) evaluations of landscapes, develops proposals regarding the most efficient (optimal) use of resources and transformation of natural conditions and the planning of natural-technical systems with particular properties; it seeks to predict the consequences of technical projects which induce changes of one sort of another in nature.

With the broadening of the range of topics studied there arose a tendency toward the development within the scope of geomorphology of separate disciplines and areas, such as the physics and chemistry of landscapes, topometry, geoqualimetry, etc.

These circumstances permit us to further consider questions of theory and methodology, grouping publications around the key words most frequently used in geomorphology, those expressing the most important concepts. First, however, we must consider the development of these same terms and concepts.

Concepts, definitions, terms. Two processes characterize the period in question. The clarification of previously formulated concepts proceeded as before. This process affected all the theoretical, methodological, and applied divisions of geomorphology and earth science. Typical were

– the subsuming of a concept under a more general one; in particular, the discussion of the relation between the natural complex and system concepts continued [247, 204, 17]

– clarification of the limits of the diffusion of a concept or term: a) through comparison of concepts with those of other sciences, such as comparison of the geomorphic and architectural-planning landscape concepts [247, 211] or between the geomorphic and biological versions of it [119, 311]; comparison of the landscape and ecosystem concepts [61, 314, 315]; b) by clarifying visible characteristics, such as the completeness of a set of components [243], by considering the boundaries of the geographical mantle as those of a physical body [236], by clarifying the content* of the concept of a facies [164]; or by replacing one visible feature by another, such as that of emergence of a biocenosis with emergence of a phytocenosis, or clarification of the concept of geomorph or landscape (e.g., [101]). There was a chaotic profusion of proposals having to do with clarification of taxonomic categories. Recommendations of this sort are contained in almost any regional geomorphology study (for example, [165]). However, most of them are not based on a substantive and logical analysis of previously formulated and proposed concepts, and so their influence on the development of the science is quite small. There still exists no unified set of definitions of members of a taxonomic series

— the terminological differentiation of concepts that designate both a process for obtaining a product and the product itself. For example, the separation of the concept of appraisal into appraising (the process) and an appraisal (the product) [145, 181].†

The broadening of the conceptual basis and enrichment of terminology were important aspects of work published during the period surveyed here. The concepts of the geosystem [247], geotechnical metabolism [186], natural aquatic complexes [244, 275], environmental contrast [172], critical geomorphic points [173], and serial-dynamic sequences [247] were actively discussed during this period. The introduction of new geomorphic classifications was proposed; for example, field of vision *(okoyem)* and belt or strip [165].

Proposals have also been advanced regarding the introduction of new terms to designate the developing divisions of geomorphology: structural-dynamic geomorphology [247], geomorphology of geosystems [247], display geomorphology [40], topometry [10], and geoqualimetry [182].

Analogous tendencies appeared in foreign work in geography (see, for example, [303], which introduces new concepts and clarifies several previously formulated ones).

The tendency noted above toward unification of terminology and improvement of the conceptual basis was in part realized in the definitions of geomorphologic terms given in the *Encyclopedic Geographical Dictionary* [184]. An especially characteristic feature of the period in question was the reaction of many workers in the field to errors in many definitions. This reaction was reflected in publications proposing formal logical methods for developing and testing concepts [167], terminological rules [103, 164], and analysis of the information content of publication titles and key words [201].

---

*Not "extent," as usually indicated by the author of this article.

---

†This process began in geography with Rodoman, who differentiated regionalization as a process from regionalization as the *result* of this process.

GENERAL PHYSICAL GEOGRAPHY

Pivotal Issues in the Study of Natural Complexes

The systems-structural and cybernetic approaches are being introduced more and more actively into geography.

The feasibility of applying them to the study of the following problems has been discussed and determined.

— large-scale natural complexes and natural geosystems [203, 204, 247, 248, 304, 187, 161]. A model of the natural complex as a self-regulating information systems has been proposed, and means of analyzing positive and negative feedback have been indicated.

The attempt to transfer concepts having to do with natural complexes to the language of a systems-structural approach must be regarded as more than simply the formal transposition of old concepts into a new language. In its time geomorphology anticipated many of the features of the study of large systems as complex, integrated units. The use of the logical models, mathematical tools, and concepts of the theory of large systems and control theory could provide new stimuli to the development of the theory of geomorphology and its methodological apparatus. The uniting of geomorphologic concepts and those of the systems approach is sharply reducing the isolation of geography from contemporary exact sciences and creating the preconditions for utilizing the methodological achievements of general systems theory and cybernetics.

— natural complex-technological system systems [204, 88, 145, 144]. It should be noted that the theoretical bases for the reclamation and transformation of nature are being clarified to a significant degree, using these approaches

— territorial biocomplexes and ecosystems, i.e., organism (or aggregate or organisms) —environment systems [191, 187, 248]

— natural complex-human organisms systems, i.e., human ecology [248]; this approach is important in the struggle against pollution of the environment

— the nature-humanity system.

It is clear that these approaches are actively being introduced not only into the study of natural complexes per se (although studies of specific territories as systems have not yet been published), but also into areas in which geomorphology and the engineering and medical-biological sciences meet.

Attempts have been made to apply the systems-structural approach to the study of the investigations processes characteristic of geomorphology [10, 181].

Unlike the situation in biology (and geobotany in particular) the systems-structural approach has not yet been applied to a sufficient degree for creating hierarchical classifications and for discovering levels of complexity and organization in natural complexes.

The introduction of the systems-structural approach into geomorphology has been accompanied by an increase in the attention devoted to modeling and the application of information approaches.

The term "structure" continues to be used in geomorphology in two senses: to characterize the links between the components of Preobrazhensky's monosystem model [204, 247, 135, 136] and to characterize the interrelations between lower-order complexes in the polysystem model [165, 113, 150].

In the area of methodology the use of data collected by research stations in the study of the structure of the monosystem model [135], and also of statistical methods there [256], information theory and statistics being used in the study of polysystems [183, 64, 65, 66, 67, 113], is of significance.

Physical-geographic, complex, and geomorphologic regionalization. In accordance with the above discussion of the use of the term, regionalization is used to designate, on the one hand, the result of the study of the differentiation of the accessible portion of the geographical mantle (a regionalization pattern) and, on the other, as the process of obtaining this result. During the period in question analytical surveys and evaluations of the current status of this issue were published not only in article [48, 49, 53, 206] but also monograph form [213, 17]. As previously, an overwhelming majority of the publications were studies of local regionalization. Many of them (including the monographs) presented results of interinstitutional studies of regionalization for agricultural purposes. Works dealing with the regionalization of the entire territory of the USSR [143, 225, 262] and individual zones or belts [74, 196] constitute a special group.

The theoretical issue discussed most actively was that of the relation between the objective and the subjective in the regionalization process [183, 234, 19, 21]. This discussion unquestionably stems from the different positions geomorphology assumes with regard to whether our consciousness, in the process of studying geographical reality, reflects its objective properties — its differentiation into relatively independent integral formations (natural complexes) or artificially constructs a representation in accordance with various research goals and problems, logical rules, and consensus among investigators. The fundamental answer given by geographers who maintain dialectical materialistic positions is clear. At the same time, this discussion also reflects the weakness of the contemporary understanding of the *relation between the boundaries of continuity and discontinuity in the geographical mantle and in natural complexes.* This permits adherents of absolute continuity, while admitting the reality of the continuity of the geographical mantle, to simultaneously deny the reality of the existence there of relatively independent integral formation (natural complexes), and to consider the introduction of them into science only the result of consensus among investigators, i.e., as a subjective abstraction. In contemporary discussion of this issue we encounter elements of both sides of this argument [19, 21]. The problem is rendered more difficult by the relative lack of

GENERAL PHYSICAL GEOGRAPHY

research into the principles governing the regionalization process treated as a specific reflection of geographical reality and into the problem of unity in physical geography and geomorphology [74].

Of great significance during the period in question on the methodological level were attempts [93, 21] to realize the proposals for the formalization and unification of investigative operations in the regionalization process given in the works of Armand [18] and Rodoman [228]. These tendencies were further developed in projects involving the use of statistical pattern — recognition algorithms on a computer for solving regionalization problems [148].

Considerable attention was also devoted to the development of methods of obtaining quantitative characterizations of homogeneity-nonhomogeneity for the results of the regionalization process [113, 66, 68, 69, 10].

There also exist here a large number of works dealing with specific regions and concretizing previously expressed theoretical propositions.

The classification, systematization, typology, and taxonomy of geomorphs. The following were significant during the period under consideration

— proposals aimed at unifying the many taxonomic grades into three levels: the planetary, the provincial, and the geomorphic [247, 303]

— attempts to find general criteria for differentiating not only territories, but also aquatories [169, 173, 226], using concepts associated with the vertical structure of complexes

— attempts to extend geomorphologic concepts to new objects: rivers [174] and glaciers [32]

— attempts to introduce a time factor into classification [247, 288];

— attempts to utilize the working tools of logic in the construction of classification schemata [166, 167, 214]

— attempts to utilize automatic codes for classification [149].

Attempts to combine individual and typological classifications [118] and to develop typology for the biological cycle [198] continued.

Attempts to analyze and discuss problems of classification within the limits of typologically homogeneous geomorphic groups were quite fruitful: for deserts [195], polessies* [1,2], forests (Ribin, 1970), and mountainous areas [120].

Dynamics and Development of Landscapes. This section also covers research into the formation of landscapes and the results of anthropogenic influence on them.

The following were the significant stages of thinking in this area: the elaboration of concepts regarding the dynamic approach — the study of dynamic and evolutionary processes in integral geographical complexes

---

*Vast alluvial plain (from the geographical name for the region of the Pripyat' River basin).

[24] and of multiply dynamic orders of natural complexes [247, 136]; the application of ideas and methods of mathematical modeling to the solution of problems involving the evolution of combinations of natural components (partial natural complexes) and landscapes (use of the principles of automatic control — 237; analogy with finite automata — 87). Most numerous were publications presenting the results of research into specific factors in the dynamics, evolution, and, consequently, the differentiation of landscapes. As previously in these publications geological and climatic factors drew the most attention. The number of investigations of the role of various anthropogenic factors is rapidly growing.

The most studied of the geological factors are tectonic structures and slow tectonic movements (lithology) [265, 34, 41].

The climatic factors that have been widely discussed include the variation of humidity [90] and thermals [212], as well as glaciation phenomena ([34] Martsinkevich 1969; [285]). Noteworthy in these publications are the frequent confusing of the concepts of evolution and change in general, the absence of *reliable objective techniques for elucidating the role (value, relative importance) of different factors and the inadequate use of new methods.*

Investigations of the role of the snow cover [129, 114, 184] in which statistical methods have lately been widely used constitute a separate category in this group of publications.

The most shaky area in research on the dynamics and development of natural complexes remains the determination of the time intervals within which homogeneous change in complexes (a departure from paleogeography and historical geography) can be observed, which hampers the use of the results obtained for the purposes of prediction. The methods and techniques for determining the absolute age of the natural objects of which landscapes are composed (radiocarbon dating, dendrochronological dating, etc.), which have already been utilized in other geographical sciences, have not yet been widely applied in geomorphology to the solution of this problem.

The investigation of cyclic and rhythmical changes in natural complexes has emerged over the last few years as a special category in the study of dynamics [224]. The basic concept in this new area is the attempt to elucidate the relation between changes in a natural complex (primarily its climatic and biological components) and periodic cosmic processes. Statistical methods are widely used for this purpose.

The study of the results and mechanisms of *anthropogenic influences on nature* constitutes a separate area in the study of the dynamics of natural complexes. Most of this work, however, is being done not on the geomorphic but on the physical-geographic level: the influence of some form of human activity on a given component is studied. Papers are appearing that deal with changes occurring in natural landmarks and groups of natural landmarks [288]. As in the preceding period, the investigation of processes associated with agricultural reclamation [150, 233] and the

influence of hydrotechnical systems [223] and occur primarily in
soils and vegetation developed extensively. A new factor in this regard
was an increase in the attention devoted to the problem of pesticides [47].
Increasing attention is also being devoted to the effects of other forms
of human activity: forestry, recreation [122, 271], transport, and ethnic
groups [84], and to the problems associated with urbanization and becoming manifest in the form of changes in the hygienic characteristics of
air and water. Little attention has been devoted to analysis of the rate at
which change is occurring, to linking this problem to that of the intensity
and form of human activity is not regarded as a regular system of operations; analysis is carried out in terms of now one, now another of isolated
aspect of it. As yet, the factual material which has been accumulated has
not permitted elucidation of significant new or general principles of a
geographical character. Clearly, new generalizations deriving from this
material could be relevant to a systems (in particular, a geomorphological)
approach to the solution of methodological problems: the development of
ways of measuring rates of change and analysis of the role of specific
types of activity in the changes occurring in a given landscape. Cooperation
with the medical–biological and economic branches of the system of the
geographical sciences should also be considerably strengthened.

A distinguishing feature of this area of research is the scantiness of
multiple regional investigations [259].

Investigation of anthropogenic changes necessitated definition of
the concept of the stability of a natural complex as symmetrical to that
of variability [208]. Attempts were made to develop methods for measuring the stability of vegetation as an important indicator of the stability of
an entire complex [122].

It is worth emphasizing that little attention has been paid to the *occurrence of natural complexes as complex unities.* On both the theoretical
and experimental levels (metabolism, energy exchange, the role of information processes, the comparative geographical study of objects on different scales), the problem of the unity of natural complexes as an independent scientific problem was not the subject of special publications during the period in question. Theoretical treatments of and publications on
the problems of prediction of changes in natural complexes [109, 189] and
the complex of natural processes [220] were few in number.

The physics of landscapes. As a new branch of knowledge, this division
of geomorphology concentrated its attention on attempts to find its place in
the study of the interaction of nature and society [19, 186, 191] through discussion of the feasibility for the control of this interaction based on an
equilibrium approach.

The clarification and elucidation of individual problems continued.

Most of these were associated with the problems of bioproductivity and, correspondingly, with agriculture [231, 82]. Publications on obtaining new physical characteristics of both individual landscapes [219, 129, 168, 81] and the geographical mantle as a whole [236] were especially numerous. Serious attention was devoted to methods of observation and measurement.

The chemistry of landscapes. In the study of the interaction of nature and society the attempt was made to broaden the range of problems studied to include, beside the traditional problems involved in the search for mineral resources (e.g., [175]) and agriculture [72], those associated with public health and human ecology [89, 137], especially the prediction of human environmental factors.

Great interest was shown in the study of natural aqueous complexes (Simonov, 1968).

There was a high proportion of regional publications. Among these an important place was occupied by attempts to arrive at typological characterizations of natural zones, morainic landscapes [138], and renaturalized landscapes [307]. The number of chemical elements studied on the geomorphic level increased [e.g., 163]. A significant number of publications relevant to this area and dealing with problems of typology, taxonomy, and regionalization [70, 72, 197, 198] also appeared.

In the area of methodology attention was concentrated on the elucidation of empirical dependencies, using methods of mathematical statistics [245, 253].

The following on the whole are significant in the area of physics and chemistry of landscapes: 1) attempts to formulate a number of new approaches to the problem of the interaction of nature and society (the problem of metabolism); 2) the expansion of the set of characterizations of the properties of natural complexes. In the area of the physics of landscapes, attempts to use new methods to solve the traditional problems of geography, formulated in terms of traditional concepts (analysis of the whole into elements), continued.

Methodological Problems

In recent years increased attention has been devoted to methodological problems and attempts have been made to interpret the methods of geomorphology [207] as reflections of a complex multistage conceptual process, the following stages being especially important:

1. observation and recording of the results of observation, involving both traditional landscape mapping and the use of research stations, which is still in the beginning stages

2. discovery of empirical dependencies; in the period in question, the

possibility of applying mathematical methods for this purpose evoked the most interest

3. development of theory; in this regard the possibilities of modeling (which is used, however, at all stages of cognition) evoked the greatest interest.

Observation and measurement. No new programs or apparatus for observation were introduced. However, based on publications presenting research data, the range of questions included in programs of observation widened disproportionately. Especially notable was the increase in the variety of measurements performed. The question of the principles governing the development of optimal observation programs [207], optimal methods for choosing the properties to be studied [147], and the best means for making comparisons between economic evaluations of exploratory and general scientific undertakings [202] were formulated.

Geographical descriptions and characterizations and the recording of observations. In the methodology of field research, along with increased mapping of observations, new nonmapping forms of recording were introduced. The most important of these was the use of punched cards [6, 271, 300]. The recommendations made in previous 5-year plans regarding the use of blank forms were widely put into effect even in textbooks during the period in question [267, 195].

The following issues were discussed in the area of geographical descriptions and characterization, i.e., the recording and generalizing of the results of empirical investigations.

— analysis of the logical structure of characterizations and ways of formalizing them [229, 102]

— comparability of characterizations [180, 103, 210]

— the historical evolution of textual characterizations [4]

— the excess of quantitative [258] and physical–chemical [264] data in geographical characterizations

— recommendations regarding the content and structure of characterizations [210, 215, 216].

Landscape mapping (cartography) and the landscape map. The overwhelming majority of publications presented information on landscape mapping projects in various regions based on principles and methods proposed previously. This accumulated experience was reflected in the manuals and textbooks used in institutions of higher education [38, 111, 112, 195, 267]. The methodology of preliminary studies improved [39], methods involving the use of aerial survey data of [231] and field maps [188] were refined, and problems associated with the mapping of the dynamics of complexes were discussed (Frisch, 1968). Researchers devoted particular attention to the methodology of applied landscape cartography: in agriculture [195, 294] and territorial planning (Arnold, 1969; [541]).

Research stations. The number of research stations increased [200], especially in Siberia [152, 135]. The first significant results of the Kursk Station (Geographical Institute of the Academy of Sciences of the USSR) [82] and the Kharanorsk Station (Siberian Geographical Institute. Siberian Division of the Academy of Sciences of the USSR) [256] were published. The papers published by the former concentrated on the study of the dependency of bioproductivity on hydroclimatic factors. Of particular interest were the comparative data on the productivity of natural complexes and agricultural areas. The geomorphologic approach was weakly developed in them. Publications resulting from the operation of the latter station, similar in direction to those of the former station, were based on a systems–structural approach and on dynamic concepts. A statistical model of the complex was used. In addition, the results of a 20-year multipurpose project carried out by the Tien-Shan High Altitude Physical Geographic Station were published [152]; this station has still not participated in the area of the systems–structural investigation of landscapes.

It appears that the relatively limited results of the activity of most research stations is due in significant measure to the absence of a precise theoretical conception of how to organize the scientific activity of a number of specialists in various disciplines for the purpose of studying one natural complex (or several) as a complex whole.

Aerial methods. Aerial methods for studying natural complexes were dealt with during the period in question primarily in textbooks [239]. No specifically geomorphological standards of interpretation were published. The results of combined studies of changes in natural elements using information obtained from one-time overflights were published [192, 193].

The extensive development of methods of extrapolating and interpolating geobotanical data were of great significance in general physical geography [42].

Observation satellites. The extensive development of satellite observation techniques and the success of their experimental application in solving meteorological problems stimulated considerable interest in such methods for studying local, regional, and planetary laws governing the composition, structure, dynamics, and rhythms of the geographical mantle through direct photographic and instrumental observation of the earth's electromagnetic field [43, 44, 45, 26]. The results of a study of the zonality of Africa (by means of photographs obtained from the Zond-5 satellite) were published. At the present time the principles and methods of interpreting the various types of data obtained from aerial instrumentation are being actively developed [44, 45].

Mathematical Methods.* Many geographers have in the last few years moved from consideration of the desirability and usefulness of the use of mathematical methods in geomorphology to rather broad experimentation in their applications in this field. At the Fifth Conference of the Geographical Society of the USSR, of the 19 papers dealing with the state and problems of physical geography, four dealt with the application of various mathematical methods [11, 114, 68, 148].

Recently, the use of quantitative characterizations has come to be identified with the use of mathematical methods. At the present time quantitative characterizations are becoming indispensible elements in most landscape descriptions. Thus, publications having titles containing such terms as qualitative characterizations, qualitative indicators, or qualitative descriptions are attracting less attention.

In publications appearing during the period in question and devoted to the general problems involved in using mathematical methods, attempts were made to determine which areas of mathematics are applicable to the solution of current problems in geomorphology on the empirical and theoretical levels [205, 24], and to formalize physical–geographic and geomorphologic problems [29, 86, 9].

Experiments in the use of mathematical methods have taken several directions. It is most appropriate to consider them in terms of the kind of apparatus used:

1. the use of mathematical statistics
2. the use of the methods of information theory and logical-informational analysis
3. the use of other mathematical methods.

It should be emphasized that much of this work was based on the use of computers.

Statistical methods were the most widely used. Their use is based on the probabilistic nature of natural complexes and processes [7, 8, 9, 10, 11, 149]. The limits of the usefulness of mathematical statistics, deriving from the incomplete correspondence of geomorphic objects or geocomplexes to the objects that mathematical statistics is classicly applied to, were discussed.

---

*This section was written by T.D. Aleksandrova.

Statistical methods are being used at the present time basically for the purpose of studying correlations, the structure of geocomplexes, and regionalization.

The study of correlations. The relatively rapid introduction of statistical methods into topography is facilitating their broad application in other geographical sciences. Thus, in earth science they are being used to ellucidate sun-Earth correlations [95, 96]. Statistical methods, especially correlation analysis, are being widely used in geomorphic-geochemical research [176, 253].

In recent years regression and dispersion analysis have come to be widely used in physical geography for investigating correlations between various phenomena [28, 153, 141, 161, 256].

The most precisely planned and executed experiment in the use of statistical methods for analyzing the interaction of the elements of a geosystem was described in the paper by Irkutsk geographers entitled "The Topology of Steppe Geosystems" [256]. The authors used correlation, regression, and dispersion analysis to establish the major dependencies in steppe geosystems. This research also involved graphical and mathematical modeling of geosystem elements (in particular, the productivity of the steppe grass stand as a function of various factors acting on it).

The structure of geocomplexes and regionalization. Kuproyanova [146, 147] has investigated the use of elements of the statistical theory of experiment design for the purpose of limiting the number of features used in regionalization studies. Mathematical statistics is being used rather actively to objectivize physico-geographical regionalization processes [110].

Experiments have been performed in the use of computerized statistical algorithms in regionalization [241, 148]. Quantitative criteria for the characterization of homogeneity-nonhomogeneity and the morphological structure of natural complexes [183, 266, 113, 65] are being sought. After examining cartographical-statistical methods of analyzing the landscape structure of physical-geographic regions, Ivashutin and Kikolayev proposed a coefficient of topographical homogeneity-nonhomogeneity that took into account the number of genetic landscape groups and the area occupied by individual groups. Gerenchuk, Gorash, and Topchiyev have proposed characterizing morphological structure by means of graphs, structural matrices, distribution curves of lower-order geocomplexes, and coefficients of complexity.

It should be noted that attempts, as yet reported only to a limited extent in the literature, are being made to use information theory for quantitative analysis of correlations between phenomena, i.e., for quantitative characterization of the diversity of the objects studied [115, 217, 68, 69]. Information theory is being applied on the largest scale in medical-geographic and bio-geographical research. Puzachenko's research is especially interesting; he has established that both two-valued and many-valued logical functions and complex logical expressions are reflected as single-valued logic in the dimensions and concepts of information theory. He proposes a means of constructing the complex logical expressions of a many-valued logic in such a way as to permit prediction of the state of a phenomenon for nonstudied combinations of factors.

The use of other mathematical methods. Devdariani's article [87] should be mentioned in this regard; it deals with the use of elements of the finite-automata theory for constructing mathematical models of physical-geographical phenomena. Also significant are the work of the brothers Sergin [237], dealing with the use of methods from automata theory (the construction of functional and structural schemata) in the analysis of the dynamics of a complex geographical object, and an article by Rakita [215], dealing with application of the "input–output" method in the study of natural complexes.

Logical methods. The possibilities offered by the methods of logic, which were first clearly formulated by Armand [18], were developed in physical geography by Milovidova [166, 167], who considered orthogonal Venn diagrams as means of comparing the calculations of various authors, and also the use of logical possibility trees for constructing and testing physical-geographic classifications. Armand [21] demonstrated the use of Euler circles in showing the interaction of phenomena. Aleksandrova [9] and Devdariani [86] attempted to use the symbols of mathematical logic to formalize the description of physical-geographic problems.

Modeling. The introduction of the concept of modeling into geomorphology occurred only during the period under consideration. The question of the construction of models of natural complexes received the most attention, [204, 308, 309, 17, 256], followed by the processes involved in their evolution [87, 237] and dynamics [248], and, finally, the process of studying landscapes for the purpose of constructing a topometrical and geoqualimetrical methodology. Ideas were formulated regarding types of models of natural complexes [204, 308, 309] and their classification [17]. The first experiment in the construction of a statistical model of a natural complex was performed [256].

Some Problems in the Theory and Methodology of Applied Geomorphological Research

The expansion of applied research has already been noted above.

As in the preceding period, research and publications were especially extensive in the area of the use of data of geomorphologic research in agriculture [170, 195, 67, 294]. Research was carried out on the reclamation of territory for industry [144, 133], transportation [195, 270], recreation [37, 271], urban construction [128, 240], and public health.

It is noteworthy that in the previously undifferentiated area of "applied geomorphologic research," the specialization of research programs, division of labor, and differentiation of the interests and knowledge of individual scientific collectives and groups of investigators have become increasingly more apparent. In our view the "application of geomorphology data," the collection of facts (cadastre-inventory data) for use in agriculture, hydroelectric construction, recreation, and public health is beginning to constitute an "engineering geomorphology," an "agro-geomorphology," a "recreation geomorphology" and a "medical geomorphology." The theoretical basis of these disciplines is the concept of the interaction of the natural complex with technical systems or an aggregate of tools interconnected with each other and the natural complex through a technological process, i.e., the concept of the interaction of a natural complex with human beings. This interaction is becoming, on the one hand, the object of engineering and agricultural manipulation, and, on the other hand, an object of study of the appropriate scientific discipline bordering geomorphology and the engineering or medical–biological sciences. These disciplines create their theoretical basis using the structural–systems approach and knowledge of the principles governing a) the natural complex, b) the engineering system (or group of human beings), and c) the interaction between them.

These trends are still in their initial phases and so publications devoted to their theoretical development are as yet few in number. Methodological literature is much more extensive. Here we wish to touch on only certain methodological processes — appraisal, planning, and indexing — which are encountered to one extent or another in most of the problems involved in the geographical analysis of almost any human activity.

Projecting and Planning. The experience accumulated in previous years by physical geographers, in particular geomorphologists, in preplanning, planning, and survey projects (specifically in the compilation of technical economic data, regional planning, the construction of State economic production plans, etc.) stimulated discussion of the question concerning the role in planning of physical–geographic research into the structure of natural complexes and landscapes [168, 116, 280, 211, 240, 272].* At the same time, increased attention was devoted by urban planners to consideration of regional specifications of natural conditions. Requirements to the effect that natural conditions should be taken into account are beginning to be included in planning manuals and methodological directives; a number of surveys on construction planning in various types of locality are being written [e.g., 76, 199, 25].

However, in most publications by geomorphologists the question of the role of research in planning has reduced to that of applied research in general, and attention has concentrated for the most part on proofs of the usefulness of the participation of geomorphologists in planning research. The number of publications in which attempts are made to treat the theoretical side of the interaction of the two scientific systems — physical geography and construction — as a tangible interaction of technical systems and natural complexes is small [203, 204, 108, 145, 147].

Today, the usefulness of the geomorphologic approach (and, in particular, of geomorphologic maps as a "subbasis" of functional zoning), as well as the participation of the geomorphologist in planning, may be considered as established. These considerations apply to elements of planning such as the collection of data on a locality, the evaluation of these data, the elucidation of territorial boundaries and regional norms, functional zoning, the engineering preparation of a territory, the selection of optimal planning alternatives, etc.

The decisive step in the development of the landscape basis of territorial planning will occur only as a result of a great effort to extend both theory and experimentation. The following will be indispensable elements in this effort:

— improvement of the systems–structural approach (to the "natural complex–engineering system" system)

— analysis of the "stability" of the natural complex, elucidation of the boundaries of the stable state under the influence of various types of human activity

— the creation of a scientifically based methodology for determining various norms and safety factors for specific complexes and types of them. Here it should be kept in mind that the use of "norms" (i.e., legally established paramenters reflecting objective, empirical dependencies between the stability of a natural complex and the behavior of the influencing system) is an indispensable feature of planning. Without norms we are deprived of the possibility of evaluating the safety margin of a natural complex, of controlling the intensity of influence on it, and of establishing a scientific

---

*Significant experience with regard to a similar range of questions has been accumulated by geographers in the German Democratic Republic, Sweden, Austria, and the Federal Republic of Germany. In the publications describing this experience emphasis is given to so-called landscape planning [A Conference on Landscape Planning, 1966: An Introduction to Landscape Planning, 1966: A Handbook of Landscape Planning, 1968: a bibliography (FRG)].

basis for measures aimed at preserving and restoring it. The scientific basis for the establishment of norms is understanding of the stability mechanism and of how to control it, the mechanism of the interaction (exchange of substances, energy, and information) between the engineering system and the natural complex

— the development of a methodology for predicting the state of a natural complex resulting from spontaneous, nonperiodic, and periodic natural changes, on the one hand, and the various forms of human activity on the other. The problem of prediction from the viewpoint of geomorphology has as yet been studied very little, either theoretically or experimentally, although geomorphologic prediction must be regarded as one of the cornerstones of territorial planning

— the creation of a methodology of full-scale investigation and modeling (including mathematical) of interaction processes (exchange of substances, energy, and information) occurring in various natural complexes and engineering systems. Much may be derived in this area from experience accumulated by the engineering branches of the physical–geographic subsidiary sciences (geocryological engineering, construction and aviation climatology, etc.), and geological engineering

— the accumulation of experience in cooperative endeavors by geomorphologists and planners at various stages of the planning process.

Appraisal of landscapes. In the past few years the character of the research done in this area has changed significantly. The introduction of elements of the systems–structural approach [145, 181, 182, 154] facilitated a clearer delineation of the subject matter of appraisal research — of the relation between natural complexes and territorial, social, engineering, and medical–biological systems, on the one hand, and analysis of the process of arriving at an appraisal as a particular form of cognitive activity [182], on the other. The beginnings of a geoqualimetry — the discipline which studies the principles and methods of measurement of the characteristics of geographical systems — were noted [182].

Together with the development of the principles and methods of appraising natural complexes from the viewpoint of agriculture, forestry, and construction, there was, as in other areas, an increase in the number of publications encompassing previously little-studied social and medical–biological aspects of human activity [35, 36].

The range of methodological questions discussed in the literature — the transition from measurement to appraisal [182], the use of statistical methods and computers in appraisal research [276], and psychological measurements — increased.

It is interesting that activity in the area of appraisal of nature facilitated the introduction of geomorphologic concepts and approaches in the English-speaking countries, in which the theoretical roots of this field previously did not exist.

Indexing (indexing geomorphology). This field uses physiognomic components as indices or decipients of landscapes [40, 48]. This important applied discipline continued during the period in question to strengthen its position in geological surveying and the search for useful minerals, including water (Vostokova *et al.,* 1968) and in the study of the engineering conditions existing in various territories (Sulakshin *et al.,* 1968; [270]). It also continued to expand into new areas, in particular the study of the dynamics of nature [193] and natural processes [40].

The concept of using the morphological structure of natural complexes as an index continued to accumulate support.

In summarizing the development of geomorphology during the period in question, it is appropriate to mention once again the unsolved problems and slowly developing tendencies, which, clearly, should be subjected to closer analysis in the near future.

The most important of these are the following
– problems of the integrity of natural complexes, the analysis of the informational and physical mechanisms involved, which have not been adequately studied
– general theoretical questions of classification, in particular the formulation of requirements that should be met in proposals for new taxons, i.e., the creation of a system of "diagnosis" for assigning observed natural complexes to one or another taxon, class, or type; these questions have not as yet received sufficient attention
– the methodology of the study of dynamics and development of natural complexes, which is progressing slowly
– the elaboration of the principles governing the study of anthropogenic factors and the development of methods for studying and classifying transformed landscapes, without which successful planning of such transformation and the creation of natural complexes with given optimal characteristics are impossible
– truly complex (and not simply multipurpose) station-based investigation of landscapes, which has received insufficient attention
– communication and cooperation with the individual physical–geographic sciences in the area of determining the characteristics of and interrelations between the components of complexes, which have been inadequate.

The planned intensification of industrial production will require an increase in the attention devoted by physical geography to regions with long-established populations. Geographers will have to deal with especially intricate natural complexes which as a rule will have been extensively altered by human activity, and occasionally transformed into controlled natural-engineering systems. Here the activity of the investigator will not be limited to preliminary surveys, mapping, and the creation of cadastres according to standard procedures, but will be a complex process, the success of which will depend on the originality of the procedures used, which in turn will depend on current theoretical knowledge, on the reliability of its realization, on the instrumentation used in the observation stage, and the efficacy of the methods employed to determine empirical regularities.

LITERATURE CITED

1. Abaturov, A.M. Topographical characteristics of polessie sand plains as related to the problem of reclamation of polessies. In: Museum of Earth Studies, Moscow State University, 1967, No. 4.
2. Abaturov, A.M. Polessies of the Russian plain in connection with the problem of reclaiming them (Poles'ya Russkoy ravniny v svyazi s problemoi ikh osvoyeniya), Moscow, Mysl' Press, 1968.
3. Abramov, L.S. Survey of work on physical-geographical area studies in the USSR from 1954 to 1964. In: Physical geography. Hydrology (Fizicheskaya geografiya. Gidrologiya), Issue 1, Moscow, 1967, pp. 8-12.
4. Abramov, L.S. Analysis of changes in the textural characteristics of natural regions. In: Methods of geomorphological research (Metody landshaftnykh issledovaniy), Moscow, Nauka Press, 1969, pp. 103-106.
5. Abramov, L.S., Armand, D.A., Nasimovich, A.A., Rakhilin, V.K. Lenin and the conservation of nature in the USSR, Isvestiya (Bulletin) of the Academy of Sciences of the USSR, Seriya Geografiya, 1970, No. 2.
6. Aleksandrova, T.D. Perfomaps in physical-geographical research, Nauka Press, 1967a, 51 pp., Ill.
7. Aleksandrova, T.D. Use of indicators of level of correlation between qualitative characteristics of components in topographical research. In: Physical geography. Hydrology (Fizicheskaya geografiya. Gidrologiya), Issue 1, Moscow, 1967b, pp. 21-23.
8. Aleksandrova, T.D. Possible ways of using statistics in topography. In: Mathematical methods in geography (Matematicheskiye metody v geografii), Moscow, Moscow State University Press, 1968, pp. 78-80.
9. Aleksandrova, T.D. Statistical methods in topography. In: Methods of topographical research (Metody landshaftnykh issledovaniy), Moscow, Nauka Press, 1969, pp. 43-70.
10. Aleksandrova, T.D. Analysis of possibilities and limitations in the application of methods of mathematical statistics in topographical research, Izvestiya (Bulletin) of the Academy of Sciences of the USSR, Seriya Geografiya, 1970a, No. 5.
11. Aleksandrova, T.D. Some general problems in the application of mathematical methods in complex physical geography. In: Materials of the Fifth Conference of the Geographical Society of the USSR (Materialy V s"ezda Geograficheskogo Obschestva SSSR), Leningrad, 1970b.
12. Alekseyev, B.A., Lukashova, Ye.N. High-altitude spectra of the Andes, Vestnik (Journal) of Moscow State University, Geography, 1969, No. 4, pp. 22-31 (English resume).
13. Alpat'yev, A.M. Hydrological cycles in nature and their transformations (Vlagooboroty v prirode i ikh preobrazovaniye), Leningrad, Gidrometeoizdat, 1969.
14. Alpatiyev, A.M. Interaction between the hydrosphere and the biosphere. In: Twenty-third Herzen Lectures, Interagency Conference. Geology and geography (XXIII Gertsenovskiye chteniya. Mezhvuzovskaya konferentsiya. Geologiya i geografiya), Leningrad, 1970.

15. Alibekov, L.A., Gerenchuk, K.I. Morphological structure of mountainous landscapes, Geomorphological collection (Landshaftnyy sbornik), Moscow, Moscow State University Press, 1970, pp. 9-27.
16. Armand, A.D. Role of feedback in the development of natural complexes. In: Problems in the development of contemporary natural science (Problemy razvitiya v sovremennom estestvoznanii), Moscow, Moscow State University Press, 1968, pp. 248-257.
17. Armand, A.D. Models in physical geography (Modeli v fizicheskoy geografii), Priroda Press, 1969, No. 5, pp. 45-53.
18. Armand, D.A. Logic of geographical classifications and regionalization schemes. In: Development and transformation of the geographical environment (Razvitiye i preobrazovaniye geograficheskoy sredy), Moscow, Nauka Press, 1964.
19. Armand, D.L. Some goals and methods of the physics of landscapes. In: Geophysics of landscapes (Geofizika landshafta), Moscow, Nauka Press, 1967, pp. 7-24.
20. Armand, D.L. Reality of the landscape. In: Methodological problems in geomorphological research (Problemy metodiki landshaftnykh issledovaniy), Summaries of papers presented at a seminar on the methodology of topographical research, Moscow, 1968b.
21. Armand, D.L. Physical geography in our time (Fizicheskaya geografiya v nashi dni), Moscow, Znaniye Press, 1968b.
22. Armand, D.L. The objective and subjective in natural regionalization, Izvestiya (Bulletin) of the Academy of Sciences of the USSR, Seriya Geografiya, 1970, No. 1, pp. 115-129.
23. Armand, D.L., Mints, A.A., Preobrazhenskiy, V.S., Richter, G.D. Complex study of natural conditions and resources in the development of physical and economic geography. In: Development of the Earth sciences in the USSR (Razvitiye nauki o Zemle v SSSR), Moscow, Nauka Press, 1967.
24. Armand, D.L., Preobrazhenskiy, V.S., Armand, A.D. Natural complexes and modern methods of studying them, Izvestiya (Bulletin) of the Academy of Sciences of the USSR, Seriya Geografiya, 1969, No. 5, 5-16.
25. Architectural planning requirements on construction in regions with dry climates. Informatsionnyy obzor. (Arkhitekturno-planirovochnyye trebovaniya k stroitel'stvu v rayonakh surovogo klimata. Informatsionnyy obzor.), Moscow, Tsentr nauchnotekhnicheskoy informatsii po grazhdanskona stroitel'stvu i arkhitekture, 1969.
26. Babkov, A.A. Possible use of satellite data in regionalization of the oceans. Materials of the Fifth Conference of the Geographical Society of the USSR (Materialy V s"ezda Geograficheskogo obshchestva SSSR), Leningrad, 1970.
27. Bagdasaryan, A.D., Davitaya, F.F., Rustamov, S.G. Development of geography in the Transcaucasian Soviet republics, Izvestiya (Bulletin) of the Academy of Sciences of the USSR, Seriya Geografiya, 1967, No. 5.
28. Belova, V.A. Use of regression, dispersion, and correlation analysis in the interpretation of results of polynological investigations for purposes of paleogeographic reconstruction, Mathematical methods in geography (Matematicheskiye metody v geografii), 1968.

29. Boychuk, V.V. and Marchenko, A.S. Background and variation of the elements of the physical-geographic environment (Fon i variatsii elementov fiziko-geograficheskoy sredy), Moscow, Nauka Press, 1968.
30. Borzenkova, I.I. Some regularities in vertical geographical zonality, Trudy (Transactions) Gosudarstvennogo geofizicheskoy observatorii, 1967, 193, 53-59.
31. Borisov, A.A. The role of Lenin's ideas in the development of geography. In: Problems of geography (Voprosy geografii), Kaliningrad, 1970.
32. Bulatov, V.I. and Revyakin, V.S. Glaciers as geomorphic complexes of the geographical mantle of the Earth, Izvestiya (Bulletin) Vsesoyuznogo Geograficheskogo Obshchestva, 1970, **102**, No. 1, 54-56.
33. Bykov, V.D. (ed.). Problems of planetary geography (Problemy planetarnoy geografii), Moscow, Moscow State University Press, 1969, 195 pp., Ill.
34. Vasil'yeva, I.L., Lyubushkina, S.G., Rodzevich, N.N. Analysis of the influence of tectonic structures on the development of plain landscapes. Geomorphological collection (Landshaftnyy sbornik), 1970, pp. 28-48.
35. Vedenin, Yu.A. and Miroshnichenko, N.N. Appraisal of natural resources for the organization of recreation, Izvestiya (Bulletin) of the Academy of Sciences of the USSR, Seriya Geografiya, 1969a, No. 4.
36. Vedenin, Yu.A. and Filippovich, L.S. Discovery and mapping of the topographical variation of natural complexes. In: Geographical problems in the organization of recreation and tourism (Geograficheskiye problemy organizatsii otdykha i turizma), Summaries of reports delivered at a working conference, Moscow, 1969b.
37. Velichko, A.A. Major climatic boundary of the pleistocene, Izvestiya (Bulletin) of the Academy of Sciences of the USSR), Seriya Geografiya, 1970, No. 3.
38. Vidina, S.A. Methodological indications for large-scale field research (Metodologicheskiye ukazaniya po polevym krupnomasshtabnym issledovaniyam), Moscow, Moscow State University Press, 1962.
39. Vidina, A.A. Methods for preliminary large-scale topographical surveying. In: Methodological problems in geomorphological research (Problemy metodiki landshaftnykh issledovaniy), Summaries of reports presented at a seminar on topographical research methods, Moscow, 1968.
40. Viktorov, V.V. Display geomorphology as one of the lines of research in contemporary geography. In: Earth Science (Zemlevedeniye), Moscow, Moscow State University Press, 1967, **7** (47), pp. 235-244.
41. Vilenkin, V.L. Some regularities in the morphostructure and dynamics of landscapes of natural zones of the left-bank of the Ukraine. In: Natural and labor resources of the left-bank of the Ukraine and their utilization. Summaries of reports (Prirodnyye i trudovyye resursy Levoberezhnoy Ukrainy i ikh ispol'zovaniya. Tezisy dokladov, Issue 4, Kharkov, 1967, pp. 57-58.
42. Vinogradov, B.V. Aerial methods for studying the vegetation of natural zones in the vicinity of Moscow and Leningrad (Aerometody izucheniya rastitel'nosti prirodnykh zon M-L), Nauka Press, 1966.
43. Vinogradov, B.V. Complex decoding of images of the Earth's surface obtained from Nimbus-1, Izvestiya (Bulletin) Vsesoyuznogo Geograficheskogo Obshchestva, 1969, **101**, Issue 3, 210-217.

44. Vinogradov, B.V. and Kondrat'yev, K.Ya. Cosmic methods in physical geography and and their use in the study of the natural resources of the Earth, Izvestiya (Bulletin) of the Academy of Sciences of the USSR, Seriya Geografiya, 1970a, No. 2.
45. Vinogradov, B.V. and Kondrat'yev, K.Ya. Geographical research with the aid of satellites, Materials of the Fifth Conference of the Geographical Society of the USSR (Materialy V s"ezda Geograficheskogo Obshchestva SSSR), Leningrad, 1970b.
46. Influence of exposure on landscapes (Vliyaniye ekspozitsii na landshafty), Uchenyye zapiski (Scientific notes) of Perm' University, No. 240, Perm', 1970.
47. Vrochinskiy, K.K., Grebenyuk, S.S., Burshteyn, A.L. Pesticide pollution of reservoirs and water supplies, Gigiyena i sanitariya, 1968, No. 11, 69-72.
48. Vyshivkin, D.D., Viktorov, S.V., Vostokova, Ye.V. Display-geographical research in the Soviet Union, Materials of the Fifth Conference of the Geographical Society of the USSR (Materialy V s"ezda Geograficheskogo obshchestva SSSR), Leningrad, 1970.
49. Gvozdetskiy, N.A. Research on physical-geographic regionalization of the USSR and its theoretical results, Vestnik (Journal) of Moscow State University, Seriya Geografiya, 1967, No. 6, 3-9.
50. Gvozdetskiy, N.A. Soviet geographical research and discoveries (Sovetskiye geograficheskiye issledovaniya i otkrytiya), Moscow, Mysl' Press, 1967b.
51. Gvozdetskiy, N.A. Role of Soviet geographical research in the solution of the theoretical problems of physical geography, Vestnik (Journal) of Moscow State University, Seriya Geografiya, 1970, No. 2.
52. Gvozdetskiy, N.A., Gerenchuk, K.I., Isachenko, A.G., Preobrazhenskiy, V.S. Current state and problems of physical geography, Materials of the Fifth Conference of the Geographical Society of the USSR (Materialy V s"ezda Geograficheskogo Obshchestva SSSR), Leningrad, 1970.
53. Gvozdetskiy, N.A. and Zhukova, V.K. Issues in the physical-geographical regionalization of the Russian lowlands, Geograficheskiy sbornik Kazanskogo universiteta, Issue 2, 1967, 70-72.
54. Gvozdetskiy, N.A., Zvonkova, T.V., Krivolutskiy, A.Ye., Solntsev, N.A. Physical-geographic economic research, Vestnik (Journal) of Moscow State University, Seriya Geografiya, 1969, No. 1, 92-94.
55. Gembel', A.V. Are the natural zones of the Earth connected? In: Twentieth Herzen Lectures. Interagency Conference. Geology and geography (20-e Gertsenovskiye chteniya. Mezhvuzovskoy konferentsii. Geografiya i geologiya), Leningrad, 1967, pp. 9-12.
56. Kostrits, I.B., Pinkhenson, D.M., Kalesnik, S.V. (ed.). Geographical Society of the USSR, 1917-1967 (Geograficheskoye Obshchestva SSSR, 1917-1967), Moscow, Mysl' Press, 1968.
57. Geophysics of landscapes (Geofizika landshafta), Academy of Sciences of the USSR, Institute of Geography, Moscow, Nauka Press, 1967.
58. Geochemistry of landscapes, Issue 3. Materials of the Moscow Branch of the Geographical Society of the USSR (Materialy Moskovskogo filiala geograficheskogo obshchestva SSSR), Moscow, 1969, 24 pp., Ill.
59. Geochemistry of landscapes, Issue 3. Materials of the Moscow Branch of the Geo-

graphical Society of the USSR (Materialy Moskovskogo filiala geograficheskogo obshchestva SSSR), Moscow, 1970, 39 pp., Ill.
60. Gerasimov, I.P. Transformation of nature and the development of geographical science in the USSR. An outline of constructional geography (Preobrazovaniye prirody i razvitiye geograficheskoy nauki v SSSR. O cherki po konstruktivnoy geografii), Moscow, Znaniye Press, 1967a.
61. Gerasimov, I.P. Soviet physical geography and its new structural trends, 1967b, **19**, No. 4, 257-262.
62. Gerasimov, I.P. Ecosystems, the biosphere and man, Izvestiya (Bulletin) of the Academy of Sciences of the USSR, Seriya Geografiya, 1969, No. 3.
63. Gerasimov, I.P. Scientific and technological progress and geography, Materials of the Fifth Conference of the Geographical Society of the USSR (Materialy V s"ezda Geograficheskogo Obshchestva SSSR), Leningrad, 1970.
64. Gerenchuk, K.I. Fundamental problems in physical geography (Osnovni problemy fizychnoi geografiy), Vishcha Shkola Press, Kiev, 1969.
65. Gerenchuk, K.I., Gorash, I.K., Topchiyev, A.G. Methods of determining certain parameters of the morphological structures of landscapes. In: Mathematical methods in geography (Matematicheskiye metody v geografii), Moscow, Moscow State University Press, 1968, 84-86.
66. Gerenchuk, K.I., Gorash, I.K., Topchiyev, A.G. A method for determining certain parameters of the morphological structure of landscapes, Izvestiya (Bulletin) of the Academy of Sciences of the USSR, Seriya Geografiya, 1969, No. 5.
67. Gerenchuk, K.I., Klimovich, P.V., Koynov, M.M., Miller, G.P. Theoretical basis of the application of geomorphic research in the solution of economic problems.
68. Gerenchuk, K.I. and Topchiyev, A.G. Parameters of the morphological structure of natural territorial complexes. In: Materials of the Fifth Conference of the Geographical Society of the USSR (Materialy V s"ezda Geograficheskogo Obshchestva SSSR), Leningrad, 1970a.
69. Gerenchuk, K.I. and Topchiyev, A.G. Informational analysis of the structure of natural complexes, Izvestiya (Bulletin) of the Academy of Sciences of the USSR, Seriya Geografiya, 1970b, No. 6.
70. Glazovskaya, M.A. Geomorphic-geochemical regionalization of the dry land areas of Earth, Vestnik (Journal) of Moscow State University, Seriya Geografiya, 1967, No. 5, 46-57.
71. Glazovskaya, M.A. Technogenesis and problems of topographical-geochemical prediction, Vestnik (Journal) of Moscow State University, Seriya Geografia, 1968, No. 1, 30-36.
72. Glazovskaya, M.A. Study of the geochemistry of landscapes for the purpose of increasing their biological productivity, Vestnik (Journal) of Moscow State University, Seriya Geografiya, 1969, No. 1, 10-19.
73. Gozhev, A.D. New map of dry land natural zones and provinces. In: Twenty-first Herzen Lectures, Interagency conference. Geography and geology (XXI Gertsenovskiye chteniya Mezhvuzovskoy konferentsii. Geografiya i geologiya), Leningrad, 1969, 12-14.
74. Gorbatskiy, G.V. Physical-geographical regionalization of the Arctic. Part 1. Matrix

tundra zones (Fiziko-geograficheskoye rayonirovaniye Arktiki. Ch. 1, Polosa materikovykh tundr), Leningrad State University, 1967.
75. Gorelova, E.M. Problem of unity in geography, Vestnik (Journal) of Leningrad State University, 1968, No. 18, 67-74 (English resume).
76. Urban construction in the Far North (Gradostroitel'stvo na Kraynem Severe), Moscow, Tsentr nauchno-tekhnicheskoy informatsii po grazhdanskomu stroitel'stvu i arkhitekture, 1969.
77. Grigor'yev, A.A. Laws of the structure and development of the geographical environment (Zakonomernosti stroyeniya i razvitiya geograficheskoy sredy), Moscow, Mysl' Press, 1966.
78. Grigor'yev, A.A. Notes on the development of general physical geography in the USSR. In: Problems in the history of physical geography in the USSR (Voprosy istorii fizicheskoy geografii v SSSR), Moscow, Nauka Press, 1970a, pp. 3-14.
79. Grigor'yev, A.A. Types of geographical environment. Selected theoretical works (Tipy geograficheskoy sredy. Izbrannyye teoreticheskiye raboty), Moscow, Mysl' Press, 1970b.
80. Grishanov, G.Ye. Natural complexes in the mountains of the Crimea. In: Natural and labor resources of the left bank of the Ukraine and their utilization (Prirodnyye i trudovyye resursy Levoberezhnoy Ukrainy i ikh ispol'zovaniya), Summaries of reports, Issue 4, Khar'kov, 1967, pp. 63-64.
81. Grin, A.M. and Kuk, Yu.V. Experimental research on the infiltration capacity of soils in the south of wooded zone b4s, Izvestiya (Bulletin) of the Academy of Sciences of the USSR, Seriya Geografiya, 1970, 6.
82. Grin, A.M., Rauner, Yu.L., Utekhin, V.D. Efficiency of utilization of radiation and moisture in forest steppe ecosystems, Izvestiya (Bulletin) of the Academy of Sciences of the USSR, Seriya Geografiya, 1970, 4.
83. Gudonite, M., Kolotiyeveskiy, A., Nymlisk, S., Yagunputmin'. Geography in the Baltic Republics, Izvestiya (Bulletin) of the Academy of Sciences of the USSR, Seriya Geografiya, 1967, No. 5.
84. Gumilev, L.N. On the anthropogenic factor in landscape formation (Landscape and Ethnos), Vestnik (Journal) of Leningrad State University, 1967, No. 24, 102-112.
85. Gumilev, L.N. Ethnos and landscape. Historical geography as ethnography, Izvestiya (Bulletin) Vsesoyuznogo Geograficheskogo Obshchestva, 1968, **100**, No. 3, 193-202.
86. Devdariani, A.S. Mathematical formulation of problems in geography. In: Mathematical methods in geography (Matematicheskiye metody v geografii), Moscow, Moscow State University, 1968, pp. 6-9.
87. Devdariani, A.S. Evolutionary orders of physical-geographic phenomena and finite automata, Izvestiya (Bulletin) of the Academy of Sciences of the USSR, Seriya Geografiya, 1969, No. 2, 115-120.
88. Devdariani, A.S. and Greisukh, V.L. Role of cybernetic methods in the study and transformation of natural complexes, Izvestiya (Bulletin) of the Academy of Sciences of the USSR, Seriya Geografiya, 1967, No. 6, 135-142.
89. Dobrovol'skiy, V.V. Geochemistry of landscapes and some problems in public health. In: Topographical geochemistry (Geokhimiya landshafta), Moscow, Nauka Press, 1967, pp. 40-53.

90. Dobrovol'skiy, V.V. Ten years of the geochemistry of landscapes at the Moscow Branch of the All-Union Geographical Society. In: Geomorphological geochemistry (Geokhimiya landshafta), Issue 4, Moscow, 1970, pp. 7-12.
91. Dolgushin, I.Yu. Role of climactic humidity variations in the evolution of landscapes in the taiga zone of Western Siberia. In: Geography and geomorphology of Asia (Geografia i geomorfologiya Azii), Moscow, 1969, pp. 205-211.
92. Doskach, A.G. and Trusov, Yu.P. Significance of Lenin's ideas regarding the development of natural science for the analysis of certain tendencies in Soviet physical geography, Izvestiya (Bulletin) of the Academy of Sciences of the USSR, Seriya Geografiya, 1970, No. 3, 19-28.
93. Drozdov, A.V. Methodological problems in the establishment of boundaries in topographical mapping. In: Methodological problems in topographical research (Problemy metodiki landshaftnykh issledovaniy), Moscow, 1968.
94. Drozdov, A.V. Some contemporary theoretical conceptions, methods, and trends of German geomorphology, Izvestiya (Bulletin) of the Academy of Sciences of the USSR, Seriya Geografiya, 1970, No. 5.
95. Druzhinin, I.P. Statistical evaluation of one type of concentration of frequencies of discontinuities in the long-term course of natural processes on Earth. In: Mathematical methods in geography (Matematicheskiye metody v geografii), Moscow, Moscow State University, 1968, pp. 93-95.
96. Druzhinin, I.P. Discontinuities in the long-term course of natural processes on Earth and sudden changes in solar activity. In: Rhythms and cycles in nature (Ritmy i tsiklichnosti v prirode), Moscow, Mysl' Press, 1970, pp. 15-50.
97. D'yakonov, K.N. Problems in predicting the effect of hydrotechnical systems on surrounding landscapes. In: Materials of the Fifth Conference of the Geographical Society of the USSR (Materialy V s"ezda Geograficheskogo Obshchestva SSSR), Leningrad, 1970a.
98. D'yakonov, K.N. Influence of reservoirs on the growth of riparian forests, Izvestiya (Bulletin) of the Academy of Sciences of the USSR, Seriya Geografiya, 1970b, No. 6.
99. Yermolayev, M.M. Boundaries and structure of geographical space, Izvestiya (Bulletin) Geograficheskogo Obshchestva, 1969, **101**, No. 5, 401-427.
100. Yesakov, V.A. Subject matter of the history of physical geography and a survey of historico-geographical research in the USSR. In: Problems of the history of physical geography in the USSR (Voprosy istorii fizicheskoy geografii v SSSR), Moscow, Nauka Press, 1970, pp. 169-188.
101. Yefremov, Yu.K. Some comments on controversial issues in general topography, Sbornik (Symposium) Muzeya zemlevedeniya, Moscow State University Collection, 1969, No. 4, pp. 81-91.
102. Yefremov, Yu.K. Tabular legends as a way of transmitting geographical information, Life of the Earth (Zhizn' Zemli), 1970, No. 6.
103. Yefremov, Yu.K. Development of a scientific terminology and the exchange of geographical information. In: Materials of the Fifth Conference of the Geographical Society of the USSR (Materialy V s"ezda Geograficheskogo Obshchestva SSSR), Leningrad, 1970.

104. Zabelin, I.M. Youth of an ancient science (Molodost' drevney nauki), Moscow Prosveshcheniye Press, 1967.
105. Zabelin, I.M. Evolution of geographical science. In: Outline of the history and theory of the development of science (Orcherki istorii i teorii razvitiya nauki), Moscow, Nauka Press, 1969, pp. 325-347.
106. Zabelin, I.M. Physical geography and the science of the future (Fizicheskaya geografiya i nauka budushchego), 2nd ed., rev., Moscow, Mysl' Press, 1970.
107. Zabirov, R.D. Results of twenty years of operation of the Tyan'-Shan high-altitude physical-geographic station. In: 50th Anniversary scientific session of the Academy of Sciences of the Kirgiz Soviet Socialist Republic (Yubileinaya nauchnogo sessiya, posvyashchennaya 50-letiyu Velikoy Oktyabrskoy Sotsial'noy revolyutsii), Academy of Sciences of the Kirkhiz SSR, Frunze, Ilim Press, 1967a pp. 239-243.
108. Zvonkova, T.V. Practical problems in physical geography, Vestnik (Journal) of Moscow State University, Geografiya, 1967, No. 5, 33-38.
109. Zvonkova, T.V. and Saushkin, Yu.G. Problems of long-term geographical prediction, Vestnik (Journal) of Moscow State University, Geografiya, 1968, No. 4, 3-11.
110. Zvorykin, K.V. and Uglov, V.A. Natural-ecological regionalization using the methods of mathematical statistics. In: Materials of the Moscow Branch of the Geographical Society of the USSR (Materialy Moskovskogo filiala Geograficheskogo Obshchestva SSSR), Issue 2, Moscow, 1968.
111. Zhuchkova, V.K. Organization and methods of complex physical-geographic research (Organizatsiya i metody kompleksnykh fiziko-geograficheskikh issledovaniy), Moscow, Moscow State University Press, 1968.
112. Ivanov, V.B. Methods of physical-geographic field research (Metodika polevykh fiziko-geograficheskikh issledovaniy), Moscow State University Press, 1968.
113. Ivashutina, L.I. and Nikolayev, V.A. Analysis of the topographical structure of physical-geographical regions, Vestnik (Journal) of Moscow State University, Geografiya, 1969, No. 4, 49-59.
114. Iveronova, M.I., Nefed'yeva, Ye.A., Yashvir, A.V. Role of the snow cover in the development of the topographical sphere. In: Materials of the Fifth Conference of the Geographical Society of the USSR (Materialy V s"ezda Geograficheskogo Obshchestva SSSR), Leningrad, 1970.
115. Izmailova, N.V. Informational measures of the diversity, interactions, and variability of features portrayed on maps. In: Mathematical methods in geography (Matematicheskiye metody v geografii), Moscow, Moscow State University Press, 1968.
116. Isachenko, A.G. Topography, architecture, and the organization of natural territories, Izvestiya (Bulletin) Vsesoyuznogo Geograficheskogo Obshchestva), 1966, **98**, No. 5.
117. Isachenko, A.G. Fifty years of Soviet topography, Izvestiya (Bulletin) Vsesoyuznogo Geograficheskogo Obshchestva, 1967, **99**, No. 5, pp. 384-387.
118. Isachenko, A.G. Systematization of the natural landscapes of the USSR. In: Soviet geographers of the Twenty-first International Geographical Congress, 1968. Summaries of reports and notices (Sovetskiye geografy XXI Mezhdunarodnogo geograficheskogo kongressu), Moscow, Nauka Press, 1968, pp. 87-88.
119. Ishankulov, M.S. and Rodionov, B.S. The interrelation of the lower taxons of forest

topography and geomorphology, Izvestiya (Bulletin) Vsesoyuznogo Geograficheskogo Obshchestva, 1969, **101**, No. 3, 250-254.
120. Kavrishvili, K.V. Systemization of mountain landscapes and aspects of their regional analysis. In: Soviet geographers of the Twenty-first International Geographical Congress, 1968 (Sovetskiye geografy XXI Mezhdunarodnogo geograficheskogo kongressu, 1968), Moscow, Nauka Press, 1968, pp. 94-96.
121. Kil'dema, K. Topography in the Estonian SSR. In: Development of geography in the Estonian SSR. 1960-1968 (O razvitii geografii v Estonskoy SSR. 1960-1968), Tallin, 1970.
122. Kazanskaya, N.S. and Kalamkarova, O.A. Study of changes in forests under the influence of recreational use. In: Geographical problems in the organization of recreation and tourism (Geograficheskiye problemy organizatsii otkykha i turizma), Moscow, 1969.
123. Kalesnik, S.V. Development of general physical geography in the USSR during the Soviet period, Izvestiya (Bulletin) Vsesoyuznogo Geograficheskogo Obshchestva, 1967, **5**.
124. Kalesnik, S.V. General geographical laws of the Earth (Obshchiye geograficheskiye zakonomernosti zemli), Moscow, Mysl' Press, 1970a.
125. Kalesnik, S.V. Significance of Lenin's ideas for Soviet geography. In: Materials of the Fifth Conference of the Geographical Society of the USSR (Materialy V s"ezda Geograficheskogo Obshchestva SSSR), Leningrad, 1970b.
126. Kalesnik, S.V., Isachenko, S.G., Nevskiy, V.V. Development of Soviet physical geography, Vestnik (Journal) of Leningrad State University, 1967, No. 18, 39-48.
127. Kalinin, G.P. Problems of global hydrology (Problemy global'noy gidrologii, Leningrad, Gidrometeoizdat, 1968.
128. Klimov, A.I. Physical geography and urban construction, Uchenyye zapiski (Scientific notes) of Gor'kiy State Pedagogical Institute, Issue 92, 1967, 36.
129. Kolomyts, E.G. Topographical and thermo-physical properties of the snow cover and a typology of the intrageomorphic units of the central taiga of the Sosvinsk region, Doklady (Reports) Instituta geografii Sibiri i Dal'nego Vostoka, Issue 13, 1966, 12-23, 82.
130. Kolomyts, E.G. Geomorphic relations as reflected in the crystalline structure of the snow cover, Doklady (Reports) Instituta geografii Sibiri i Dal'nego Vostoka, Issue 19, 1968, 63-74.
131. Konovalenko, Z.P. and Druzhinin, I.P. Statistical methods for analyzing cyclicity in natural processes. In: Mathematical methods in geography (Matematicheskiye metody v geografii), Moscow, Moscow State University, 1968, pp. 173-175.
132. Kornilov, B.A. and Mukhina, L.I. Geographical prerequisites for the selection of regions for economic reclamation in the Central Ob littoral, Izvestiya (Bulletin) of the Academy of Sciences of the USSR, Seriya Geografiya, 1968, No. 1.
133. Kornilov, B.A. and Mukhina, L.I. Reclamation in the taiga zone of Western Siberia, Izvestiya (Bulletin) of the Academy of Sciences of the USSR, Seriya Geografiya, 1969, No. 4.
134. Kotlyakov, V.M. The snow cover and glaciers (Snezhnyi pokrov i ledniki), Gidrometeoizdat, Leningrad, 1968.

135. Krauklis, A.A. Stationary investigation of geomorphic structure, Doklady (Reports) Instituta geografii Sibiri i Dal'nego Vostoka, 1967, No. 16, 32-41, 96 (French resume).
136. Krauklis, A.A. Factorial-dynamic orders of taiga geosystems and the principles of their structure, Doklady (Reports) Instituta geografii Sibiri i Dal'nego Vostoka, Issue 22, 1969, 15-25, 88 (French resume).
137. Krepkogorskiy, L.N. Hygienic significance of the fluorous geochemical landscape based on the landscapes of Central Kazakhstan, Trudy (Transactions) Kazanskago instituta usovershenstvennykh vrachey imeni V.I., Lenina, 1969, **24**, 2-25.
138. Krym, I.Ya. Geographical characteristics of a moraine landscape. In: The Northwest European part of the USSR (Severo-Zapad Evropeyskoy chasti SSSR), Issue 7, Leningrad, Leningrad State University, pp. 129-142.
139. Kryuchkov, V.V. High-altitude zonality in the Khibin Mountains, Sbornik (Collection) muzeya zemlevedeniya, Moscow State University, No. 4, 1969, pp. 124-131.
140. Kruchkov, V.V. Absence of forests in the tundra zone of Northeastern Siberia and its causes, Izvestiya (Bulletin) of the Academy of Sciences of the USSR, Seriya Geografiya, 1967, No. 4, 94-103.
141. Kugelevichus, I.B. and Liopo, T.N. Elucidation of primary factors in complex interrelationships in geographical research using mathematical statistics, Doklady (Reports) Instituta geografii Sibiri i Dal'nego Vostoka, 1967, Issue 15, 41-47 (French resume).
142. Kuznetsov, P.S. Methodological problems of physical geography. Selected lectures (Metodologicheskiye problemy fizicheskoy geografii. Izbrannyye lektsii), Saratov University, 1970.
143. Kuzyakin, A.P. Zonal-geomorphic regionalization of the USSR. In: Physical geography. Hydrology (Fizicheskaya geografiya. Gidrologiya), Issue 1, Moscow, 1967, pp. 18-21.
144. Kunitsyn, L.F. Reclamation of Western Siberia and problems in the interaction of natural complexes and technical systems, Izvestiya (Bulletin) of the Academy of Sciences of the USSR, Seriya Geografiya, 1970, 1.
145. Kunitsyn, L.F., Mukhina, L.I., Preobrazhenskiy, V.S. Some general problems in the technological appraisal of natural complexes for purposes of land reclamation, Izvestiya (Bulletin) of the Academy of Sciences of the USSR, Seriya Geografiya, 1969, No. 1, 38-49.
146. Kupriyanova, T.P. Statistical gaps in the limitation of the number of features analyzed in geomorphological research. In: Mathematical methods in geography (Matematicheskiye metody v geografii), Moscow, Moscow State University, 1968, pp. 80-83.
147. Kupriyanova, T.P. Application of elements of statistical theory on planning in experiment in physical-geographic research. In: Methods of geomorphological research (Metody landshaftnykh issledovaniy), Moscow, Nauka Press, 1969, pp. 35-42.
148. Kupriyanova, T.P. Statistical method of differentiating homogeneous physical-geographic regions. Record of a presentation at the Fifth Conference of the Geographical Society of the USSR of a paper by P.A. Gvozdetskiy, K.I. Gerachuka,

A.G. Isachenko, V.S. Preobrazhenskiy: The current state of physical geography, Leningrad, 1970.
149. Kuptsov, V.I. and Simonov, Yu.G. Substantiation of probabalistic-statistical conceptions in geography, Vestnik (Journal) of Moscow State University, Seriya Geografiya, 1970, No. 1.
150. Kurakova, L.I. and Ryabchikov, A.M. Reclamation and alteration of natural landscapes, Vestnik (Journal) of Moscow State University, Seriya Geografiya, 1967, No. 5, 58-67.
151. Geomorphological collection (Landshaftnyy sbornik), Moscow, Moscow State University, 1970.
152. Lashchinskiy, N.N. Physical-geographic conditions of the Opokskiy Research Station. In: Structural dynamic characteristics of the phytocenoses of the Lower Angara region (Strukturno-dinamicheskiye osobennosti fitotsenozov Nizhnego Priangar'ya), Novisibirsk, Nauka Press, 1969, pp. 5-12.
153. Liopo, N.V. Several ways of performing a quantitive appraisal of the natural processes of the geographical environment. In: Mathematical methods in geography (Matematicheskiye metody v geografii), Moscow, Moscow State University, 1968, pp. 74-76.
154. Lopatina, Ye.B., Mints, A.A., Mukhina, L.I., Nazarevskiy, O.R., Preobrazhenskiy, V.S. Current status of appraisal of natural conditions and resources, Izvestiya (Bulletin) of the Academy of Sciences of the USSR, Seriya Geografiya, 1970, No. 4, 45-54.
155. Lymarev, V.I. Application of the topographical-zone principle to the regionalization of coastal areas. In: Summaries of papers read at the 22nd Scientific Conference, in honor of the 50th Anniversary of the Soviet State (Tezisy dokladov XXII Nauchnoy Konferentsii, posvashchennoy 50-letiyu sovetskogo gosudarstva Dal'nevostochnyy universitet), Far-Eastern University, Part 2, Vladivostok, 1967, pp. 133-136.
156. Lyamin, V.S. and Samoylov, L.N. Topogenesis – an important stage in the development of the nature of the Earth. In: Problems of development and contemporary natural science (Problemy razvitiya i sovremennoye estestvoznaniye), Moscow, Moscow State University, 1969, pp. 258-275.
157. Marinich, A.I., Palamarchuk, M.M., Shcherban', M.I. Development of the geographical sciences in the Ukrainian SSR during the Soviet era, Izvestiya (Bulletin) of the Academy of Sciences of the USSR, Seriya Geografiya, 1967, No. 5.
158. Markov, K.K. Natural unity of the oceans and continents, Izvestiya (Bulletin) Vsesoyuznogo Geograficheskogo Obshchestva, 1968, **100**, No. 6, 481, 487.
159. Markov, K.K. Stages in the development of the study of world geographical zonality in the Soviet Union. In: Problems of planetary geography (Problemy planetarnoy geografii), Moscow, Moscow State University, 1969, pp. 47-61.
160. Markov, K.K. Geography of the oceans. In: Materials of the Fifth Conference of the Geographical Society of the USSR (Materialy V s"ezda Geograficheskogo obshchestva SSSR), Leningrad, 1970.
161. Mart'yanova, G.N. Quantitative appraisal of radiation equilibrium in geographical steppe environments. In: Mathematical methods in geography (Matematicheskiye

metody v geografii), Moscow, Moscow State University, 1968, pp. 88-91.
162. Methods of geomorphological research (Metody landshaftnykh issledovaniy), Institute of Geography of the Academy of Sciences of the USSR, Moscow, Nauka Press, 1969.
163. Microelements in the landscapes of the Soviet Union (Mikroelementy v landshaftakh Sovetskogo Soyuza), Glazovskaya, M.A. (ed.), Moscow, Moscow State University, 1969, 248 pp., Ill.
164. Milkina, L.I. Content of the "facies" concept. In: Geomorphological collection (Landshaftnyy sbornik), Moscow, Moscow State University, 1970, pp. 182-188.
165. Miller, G.P. Characteristics of the topographical structure of mountains, Vestnik (Journal) of Moscow State University, Geografiya, 1968, No. 3, 88-92 (English resume).
166. Milovidova, N.V. Use of logical possibility trees in the construction and checking of physical-geographic classifications, Izvestiya (Bulletin) of the Academy of Sciences of the USSR, Seriya Geografiya, 1969, No. 4, 154-159.
167. Milovidova, N.V. Application of the methods of logic in analysis of the definition of physical geography. In: Methods of geomorphological research (Metody landshaftnykh issledovany), Moscow, Nauka Press, 1969b, pp. 127-143.
168. Mil'kov, F.N. Geomorphological geography and problems of practice (Landshaftnaya geografiya i voprosy praktiki), Moscow, Mysl' Press, 1966.
169. Mil'kov, F.N. Basic problems of physical geography (Osnovnyye problemy fizicheskoy geografii), Moscow, Vysshaya Shkola Press, 1967a.
170. Mil'kov, F.N. Maps of types of locality and their practical significance for agricultural purposes. In: Geographical collection of Kazan University (Geograficheskiy sbornik kazanskogo unversiteta), Issue 2, 1967b, 73-77.
171. Mil'kov, F.N. Division of the topographical sphere of the Earth into sections and classes of landscapes. In: Physical geography (Zemlevedenie), 7, Moscow, Moscow State University, 1967c, pp. 9-16.
172. Mil'kov, F.N. Contrastability of environments and geographical consequences. In: Philosophy and natural science (Filosofiya i estestvoznaniye), 1968, Issue 2, Voronezh.
173. Mil'kov, F.N. The geomorphological sphere of the Earth (Landshaftnaya sfera Zemli), Mysl' Press, 1970.
174. Mil'kov, F.N. and Fedotov, V.I. Delineating natural landmarks in the upper Don river bed, Nauchnyye zapiski (Scientific notes) Voronezhskogo otdela Geograficheskogo obshchestva SSSR, Voronezh, 1967, 26-29.
175. Mikhaylov, I.S., Mikhaylova, R.P., Solntseva, N.P. Compilation of large-scale geomorphic and geochemical maps of mountain taiga regions for purposes of searching for mineral deposits. In: Soil geography and the geochemistry of landscapes (Geografiya pochv i geokhimiya landshaftov), Moscow, Moscow State University, 1967, pp. 135-167.
176. Mikhaylov, I.S. and Solntseva, N.P. Correlation analysis in topographical-geochemical research. In: Geochemistry of landscapes (Geokhimiya landshafta), Issues 1-2, Moscow, 1968, 14-15.
177. Mikhaylov, N.I. Physical-geographic regionalization. In: Results in science. Geog-

raphy of the USSR (Itogi nauki, Geografiya SSSR), Moscow, 1967, Issue 4.
178. Mikhaylov, N.I. Factors in spacial physical-geographic differentiation, Informatsionnyy byulleten', Nuchnogo soveta po kompleksnomu osvoyeniyu tayezhnykh territoriy. Sibirskiy otdel AN SSSR, 1969, No. 2, 114-119.
179. Murzayev, E.M., Agakhanyants, O.Ye., Babaev, A.G., Babushkin, L.N., Otorbayev, K.O. Geography in the republics of Central Asia, Izvestiya (Bulletin) of the Academy of Sciences of the USSR, Seriya Geografiya, 1967, 6.
180. Mukhina, L.I. Determining the comparability of regional physical-geographic characterizations. In: Methods of geomorphological research (Metody landshaftnykh issledovaniy), Moscow, Nauka Press, 1969a, pp. 96-102.
181. Mukhina, L.I. Methods of industrial appraisal of natural complexes. In: Methods of geomorphological research (Metody landshaftnykh issledovaniy), Moscow, Nauka Press, 1969b, pp. 79-87.
182. Mukhina, L.I. Problems in the methodology of appraising natural complexes, Izvestiya (Bulletin) of the Academy of Sciences of the USSR, Seriya Geografiya, 1970, 6.
183. Mukhina, L.I., Preobrazhenskiy, V.S., Fadeyeva, N.V. Objective prerequisites of certain methods of geomorphic regionalization, Izvestiya (Bulletin) of the Academy of Sciences of the USSR, Seriya Geografiya, 1968, No. 4, 23-31.
184. Neklyukova, N.P. General physical geography. A textbook for pedagogical institutes (Obshcheye zemlevedeniye. Uchebnik posobiye dlya pedalogicheskikh institutov), Prosveshcheniye Press, 1967.
185. Nefed'yeva, Ye.A. Approaches to the study of the influence of the snow cover on the structure of natural complexes, Zapiski Zabaikal'skogo filiak Geografiya Obshchestva SSSR, 1970, Issue 40, 92-93.
186. Neef, E. Exchange of substances between society and nature as a geographical problem. Izvestiya (Bulletin) of the Academy of Sciences of the USSR, Seriya Geografiya, 1969, No. 1, 125-135.
187. Neef, E. Problems in the comparative ecology of landscapes, Dokladi (Reports) Instituta geografii Sibiri i Dal'nogo Vostoka, 1968, Issue 19, 44-53.
188. Nikolayev, V.A. Method of small-scale cartography. Vestnik (Journal) of Moscow State University, Seriya Geografiya, 1970, No. 3.
189. Nikol'skaya, V.V. Physical-geographic prediction of the development of the Far East. In: Physical geography (Fizicheskaya geografiya), Moscow, 1969, Issue 3, pp. 16-18.
190. Markov, K.K. (ed.). General physical geography. Texts of lectures (Obshchaya fizicheskaya geografiya. Teksty lektsii), Moscow, 1967.
191. Odum, E. Ecology (Ekologiya), tr. from the English, Moscow, Prosveshcheniye Press, 1968.
192. Orlov, V.I. Development of the forest-swamp zone of Western Siberia, Leningrad, Nedra Press, 1968.
193. Orlov, V.I. Foundations of dynamic geography (Osnovy dinamicheskoy geografii), Moscow, Prosveshcheniye Press, 1969.
194. Osokin, I.M. Winter studies, its foundations, subject matter, methods, and research procedures. In: Problems of regional winter studies (Problemy regional'nogo zimovedeniya), Chita, 1968, Issue 2, pp. 3-24.

195. Pashkang, K.V., Vasiliyev, I.V., Lapkina, N.A., Rychagov, G.I. Complex field procedures in physical geography, Moscow, Vysshaya Skola Press, 1969.
196. Petrov, M.P. Classification of the Earth's deserts, Izvestiya (Bulletin) Vsesoyuznogo Geograficheskogo Obshchestva, 1969, **101**, No. 6, 489-497.
197. Perel'man, A.I. Current state of landscape geochemistry and problems for future research. In: Landscape geochemistry (Geokhimiya landshafta), Moscow, Nauka Press, 1967, pp. 5-39.
198. Perel'man, A.I. Principles for distinguishing types of geochemical landscape and landscape periodicit, Vestnik (Journal) of Moscow State University. Geografiya, 1970, No. 1, 20-27.
199. Planning and construction of populated areas in sand desert conditions (Planirovka i zastroyka naselennykh mest v usloviyakh peschanoy pustyni), Moscow, Tsentr. nauchno-tekhnicheskogo informatsii po grazhdanskomy stroitel'stu i arkhitekture, 1968.
200. Progrebnyak, P.S. Need for the construction of complex geographical research stations. In: Landscape geophysics (Geofizicheskaya landshafta), Moscow, Nauka Press, 1967, pp. 53-56.
201. Polushkin, V.A. Titles of scientific articles in geography (their information content, the correspondence between elements of the content of a scientific publication, their use in searching for information, and standardization of format), Geograficheskiy sbornik, Collection of the All-Union Institute of Scientific and Technical Information, 1969, **3**, pp. 111-117.
202. Popovichev, Ye.A. Organization of technical-economic field calculations during geomorphological engineering research in mountain taiga, Nauchnyy trudy (Scientific works) Obninskiy otdel' Geograficheskogo obshchestva SSSR, 1968, Part 1, pp. 44-53.
203. Preobrazhenskiy, V.S. Topographical research (Landshaftnyye issledovaniya), Moscow, Nauka Press, 1966.
204. Preobrazhenskiy, V.S. New landmarks in Soviet physical geography, Priroda Press, 1967, No. 8, pp. 51-59.
205. Preobrazhenskiy, V.S. Mathematical methods in the development of geomorphology. In: Mathematical methods in geography (Matematicheskiye metody v geografii), Moscow, Moscow State University Press, 1968a, pp. 17-19.
206. Preobrazhenskiy, V.S. Sources and prospects for the solution of problems of natural regionalization, Izvestiya (Bulletin) of the Academy of Sciences of the USSR, Seriya Geografiya, 1968b, No. 1.
207. Preobrazhenskiy, V.S. Methods of general physical geography. In: Methods of geomorphological research (Metody landshaftnykh issledovaniy), Moscow, Nauka Press, 1969a, pp. 7-34.
208. Preobrazhenskiy, V.S. Primary geographical aspects and problems of the organization of recreation. In: Geographical problems in the organization of recreation and tourism (Geograficheskiye problemy organizatsii otdykha i turizma), Summaries of reports presented at a working conference, Moscow, 1969b.
209. Preobazhenskiy, V.S., Abramov, L.S., Fradkin, N.G. Lenin and the organization of the study of nature in the USSR in the interests of the national economy, Izvestiya

(Bulletin) of the Academy of Sciences of the USSR, Seriya Geografiya, 1970, No. 2.
210. Preobrazhenskiy, V.S. and Richter, G.D. Methods of working in the field of natural regionalization. In: Methods of geomorphological research (Metody landshaftnykh issledovaniy), Moscow, Nauka Press, 1969, pp. 88-95.
211. Preobrazhenskiy, V.S., Chalaya, I.P., Sheffer, Ye.G. Current problems in the planning of recreation areas. In: For the urban construction planner. Recreation areas and the planting of trees in cities (V pomoshch' proyektirovshchiku-gradostroitelyu. Mesta otdykha i ozeleneniye gorodov), Kiev, 1969, Issue 5, 3-6.
212. Proka, V.Ye. Role of the climactic factor in the development of landscapes. In: Problems in the geography of Moldavia (Problemy geografii Moldavii), 2nd ed., Kishinev, Academy of Sciences of the Moldavian SSR, 1967, pp. 33-54.
213. Prokayev, V.I. Basic methods of physical-geographic regionalization (Osnovy metodiki fiziko-geograficheskogo rayionorovaniya), Leningrad, Nauka Press, 1967.
214. Prokayev, V.I. Nature and classification of geocomplexes, Uchennyye zapiski (Scientific notes) of Sverdlovsk State Pedogogical Institute, 1969a, Issue 6, 3-35.
215. Some problems in the textural characterization of elements in the physical-geographic regionalization of dry land, Uchennyye zapiski (Scientific notes) of Sverdlovsk State Pedogogical Institute, 1969b, Issue 6, 50-77.
216. Prokayev, V.I. Structure of physical-geographic characterizations. In: Materials of the Fifth Conference of the Geographical Society of the USSR (Materialy V s"ezda SSSR, Leningrad, 1970.
217. Puzachenko, Yu.G. Study of the organization of biogeocentric systems. Author's abstract of dissertation written in candidacy for the degree of doctor of geographical sciences, Moscow, 1971.
218. Rakita, S.A. Mathematical descriptions of geographical systems. In: Current problems in the reclamation of the Soviet North (Sovremennyye problemy osvoyeniya Severa SSSR), Moscow, 1970.
219. Rauner, Yu.L., Ananiyev, L.M., Rudnev, N.I. Radiation and thermal equilibrium of the primary natural landscapes of the forest steppe. In: Landscape geophysics (Geogizilea landshafta), Moscow, Nauka Press, 1967, pp. 25-39.
220. Reznikov, A.P., Kukushkina, V.P., Druzhinin, I.P. Application of informational instructional systems for prediction of natural processes. In: Mathematical methods in geography (Matematicheskiye metody v geografii), Moscow, Moscow State University, 1968, pp. 175-177.
221. Topography of the Earth. Morphostructure and Mophosculpture (Rel'yef zemli. Morfostruktura i morfoskul'ptura), Moscow, Nauka Press, 1967.
222. Reteyum, A.Yu. Definition of the geosystem concept. In: Materials of the Fifth Conference of the Geographical Society of the USSR (Materialy V s"ezda Geograficheskogo obshchestva SSSR), Leningrad, 1970.
223. Reteyum, A.Yu., Vendrov, S.L., D'yakonov, K.N. Influence of reservoirs on the climate of shore regions in various geographical zones. In: Influence of reservoirs in forest zones on adjacent territories (Vliyaniye vodokhranilishch lesnoy zony na prilegayushchiye territorii), Moscow, 1970.
224. Rhythms and cyclicity in nature, Nauchny sbornik (Scientific collection) Moskovskogo filiala Geograficheskogo Obshchestva SSSR, Voprosy geografii (Problems in

geography), Collection 79, Moscow, Mysl' Press, 1970.
225. Richter, G.D. Primary factors and regularities in the territorial differentiation of nature in the USSR and physical-geographic regionalization. In: Physical geography (Zemlevedeniye), Moscow, Moscow State University, 1969a, pp. 24-37.
226. Richter, G.D. Natural territorial complexes of the Earth, Izvestiya (Bulletin) of the Academy of Sciences of the USSR, Seriya Geografiya, 1969b, No. 5, 17-20.
227. Rodin, L.Ye. and Bazilevich, N.I. Dynamics of organic substances and the biological cycle of ash elements and nitrogen in the primary global vegetation types (Kinamika organicheskogo veshchestva i biologicheskiy drugovorot zol'nykh elementov i azota v osnovnykh tipakh rasvitel'nosti zemnogo shara), Moscow-Leningrad, Nauka Press, 1965.
228. Rodoman, B.B. Regionalization as a means of geographical characterization, its logical forms and representation in map form, Moscow State University, 1966.
229. Rodoman, B.B. Mathematical aspects of the formalization of geographical characterizations by region, Vestnik (Journal) of Moscow State University, Seriya Geografiya, No. 2, 1967.
230. Rodoman, B.B. Zonality and geographical zones, Vestnik (Journal) of Moscow State University, Seriya Geografiya, 1968, No. 5, 33-40.
231. Romanova, Ye.A. Some morphological characterizations of oligotrophic swamp landscapes of the West-Siberian lowlands as a basis for their typology and regionalization. In: Nature of swamps and methods for investigating them (Priroda bolot i metody ikh issledovaniy), Leningrad, Nauka Press, 63-67.
232. Ryabchikov, A.M. Hydrothermic conditions and the productivity of phytomasses in the primary geomorphic zones, Vestnik (Journal) of Moscow State University, Geografiya, 1968, No. 5, 41-48.
233. Ryabchikov, A.M. The anthropogenic factor in the change in the geosphere, Vestnik (Journal) of Moscow State University, Geografiya, 1970, No. 2, 90-96.
234. Sadovsky, P.S. Principles and features of a natural region. In: Problems in the geography of the Southern Urals (Voprosy geografii Yuzhnogo Urala), 2nd ed., Chelyabinsk, 1968, pp. 3-19.
235. Sakharaova, O.D. Interaction and interdependence of geomorphic complexes. The example of the Alay Valley of the Kirgiz SSR (Kyrgizskiy geograficheskiy koomunun kabarlaly), Izvestiya (Bulletin), Kirgizskiy geograficheskiy obshchestva, 1970, Issue 8, 39-43.
236. Svatkov, N.M. The subject of external boundaries and the nature of the geographical mantle (O predmete issledovaniya fizicheskoy geografii), Moscow, Mysl' Press, 1970.
237. Sergin, V.Ya., Sergin, S.Y. Mathematical description of geographical objects. First Conference on Mathematical Methods in Geography (Matematicheskiye metody v geografii).
238. Simonov, A.I. Problems in the hydrochemistry of the southern seas of the USSR, Nauchnyy trudy (Scientific works) Obninsky otdel' Geograficheskogo Obshchestvo SSSR, Collection 1, Part 1, 1968, 85-98.
239. Smirnov, L.Ye. Theoretical foundations and methods of geographical decoding of aerial photographs (Teoreticheskiye osnovy i metody geograficheskogo deshifriro-

vaniya aerosnimkov), Leningrad State University Press, 1967.
240. Smirnova, Ye.D. Geomorphology and architecture, Vestnik (Journal) of Moscow State University, Geografiya, 1969, No. 1, 3-9.
241. Soviet geomorphology, 1917-1967. A bibliographic guide to the literature (Sovetskoye landshaftovedeniye, 1917-1967. Bibliograficheskiy ukazatel' literatury), Lvov, 1970.
242. Sonechkin, D.M. Methods of pattern recognition theory in the objective classification of geographical phenomena and situations (regionalization) using computers, Matematicheskiye metody v geografii, 1968.
243. Solntsev, N.A. Theory of natural complexes, Vestnik (Journal) of Moscow State University, Geografiya, 1968, No. 3, 14-27.
244. Solntsev, N.A. Natural aquatic complexes of the world's oceans, Vestnik (Journal) of Moscow State University, Geografiya, 1969, No. 3, 20-26.
245. Solntsev, N.A. Current problems in geomorphology. In: Geomorphological collection (Landshaftnyy sbornik), Moscow, Moscow State University, 1970, pp. 4-8.
246. Solntseva, N.P. Correlation analysis in the study of the geochemistry of mountain taiga landscapes in a normal geochemical field and tungsten ore finds. In: Microelements in the landscapes of the Soviet Union (Microelementy v landshaftakh Sovetskogo Soyuza), Moscow, Moscow State University, 1969, pp. 204-247.
247. Sochava, V.B. Structural-dynamic geomorphology and the geographical problems of the future, Doklady (Reports) Instituta Geografii Sibiri i Dal'nogo Vostoka, 1967, No. 16, 18-31, 95-96.
248. Sochava, V.B. Geography and ecology. In: Materials of the Fifth Conference of the Geographical Society of the USSR (Materialy V s"ezda Geograficheskogo Obshchestva SSSR), Leningrad, 1970.
249. Sochava, V.B. Influence of Lenin's ideas on the development of geographical science, Doklady (Reports) Instituta Geografii Sibiri i Dal'nogo Vostoka, No. 26, 1970.
250. Suyetova, I.A. Areas of the geographical zones of the continents and oceans of the Earth, Doklady (Reports) of the Academy of Sciences of the USSR, 1970, **192**, No. 1, 193-195.
251. Starikov, K.Z. Schematic modelling of physical-geographic processes. In: Mathematical methods in geography (Matematicheskiye metody v geografii), Moscow, Moscow State University, 1968, pp. 76-78.
252. Stemanov, V.N. Planetary processes and natural changes in the Earth (Planetarnyye protsessy i izmeneniya prirody Zemli), Moscow, Znaniye Press, 1970.
253. Tarnovsky, A.A. Some possible uses of spectral emission analysis and of mathematical methods of statistics in geochemical investigations of landscapes. In: The Northwest European portion of the USSR (Severo-Zapad Yevropeyskogo chasti SSSR), Leningrad State University, 1969, Issue 7L, pp. 5-26.
254. Tikhonova, T.S. Influence of the exposure of slopes on the Dzhungar Alatau geomorphic zone, Vestnik (Journal) of Moscow State University, Geografiya, 1967a, No. 4, 91-93.
255. Tikhonova, T.S. Characteristics of the Dzhungar Alatau high-altitude zone, Vestnik (Journal) of Moscow State University, Geografiya, 1967b, No. 6, 96-102.

256. Topology of steppe geosystems (Topologiya stepnykh geosistem), Moscow-Leningrad, Nauka Press, 1970.
257. Fadeyeva, N.V. Survey of geomorphological research during the period of 1954 to 1964. In: Physical geography. Hydrogeology (Fizicheskaya geografiya. Gidrogeologiya), Moscow, 1967, Issue 1, pp. 5-8.
258. Fedina, A.Ye. Quantitative characterizations of physical-geographic complexes (as illustrated by the Northeastern Caucasus), Vestnik (Journal) of Moscow State University, Geografiya, 1968a, No. 1, 57-60.
259. Fedina, A.Ye. Influence of human activity on the natural environment of the Northeastern Caucasus, Izvestiya (Bulletin) Vsesoyuznogo Geograficheskogo Obshchestva, 1968b, **100**, No. 4, 337-340.
260. Fedina, A.Ye. Research on quantitative methods in complex physical geography, Elmi. Azerb. univ. Keol. chorg. elmleri ser., Uchennyye zapiski (Scientific Notes) azerb. un-t. Ser. geol.-geogr. n., 1968b, No. 4, 57-64.
261. Fedorovich, B.A. and Pal'gov, N.N. Achievements of Soviet geography in Kazakhstan, Izvestiya (Bulletin) of the Academy of Sciences of the USSR, Seriya Geografiya, 1967, No. 6.
262. Gvozdetskiy, N.A. (ed.). Physical-geographic regionalization. Characterization of regional units (Fiziko-geograficheskoye rayonirovaniye. Kharakteristika regional'nykh edinits), Moscow State University Press, 1968.
263. Fradkin, N.G. Geographical discoveries, their objects and nature at various stages in the scientific investigation of the Earth, Izvestiya (Bulletin) of the Academy of Sciences of the USSR, Seriya Geografiya, 1968, **1**.
264. Frisch, E.V. Use of physical-chemical data in the characterization of geomorphic structures, Doklady (Reports) Instituta Geografii Sibiri i Dal'nogo Vostoka, 1967, Issue 14, (French resume), 59-69.
265. Khomentovskiy, A.S. Role of geological factors in the development of recent geocomplexes, Informatsionnyy byulleten' nauchnogo soveta po kompleksnomy osvoyeniyu taezhnogo territorii, Sibirian Branch of the Academy of Sciences of the USSR, 1969, No. 2, 120-123.
266. Tsesel'chuk, Yu.N. Natural-agroproductive nonhomogeneity of arable land, Vestnik (Journal) of Moscow State University, Seriya Geografiya, 1969, No. 3.
267. Chochia, S.L. Summer field practice in geomorphology (Letnyaya polevaya praktika po landshaftovedeniyu), Leningrad State University, 1969.
268. Chukreyev, V.K. Indicators of the distribution of heat and moisture on the Earth. In: Biogeography (Biogeografiya), Moscow, 1970, Issue 4, 20-21.
269. Chupakhin, V.M. High-altitude zonality as a basic regularity in the geomorphic differentiation of mountains, and its interrelations with latitudinal zonality and regional characteristics. In: Geographical research in Kazakhstan (Geograficheskiye issledovaniy v Kazakhstane), Alma Ata, 1968.
270. Shevchenko, L.A. Possible use of geomorphological indicators in the approximate evaluation of the conditions of travel on swamps, Byulleten' Moskovskogo obshchestva ispytaniya prirody, Otdel' geologiya, 1968, **43**, No. 6, 149-150.
271. Sheffer, Ye.G. Some methods for recording the results of geomorphological research.

In: Methodological problems in geomorphological research (Problemy metodiki landshaftnykh issledovaniy), Moscow, 1968.
272. Sheffer, Ye.G. Defining the resistance factor of a natural complex to recreational loads. In: Geographical problems in the organization of recreation and tourism (Geograficheskiye problemy organizatsii otdykha i turizm), Moscow, 1969a.
273. Sheffer, Ye.G. Geomorphological research for planning purposes. In: Summaries of papers presented at the Conference of Young Scholars in honor of the 150th anniversary of Leningrad University and the 50th anniversary of the Geography Department, 1969 (Tezisy doklady k Konferentsii molodykh uchenykh, posvyashchenniye 150-letiyu Leningradskogo Universiteta i 50-letiyu Geograficheskogo Fakultata, 1969), Leningrad, 1969b, pp. 4-6.
274. Sheshel'gis, K. Geomorphological terminology in urban construction and regional planning, Nauchnyye trudy (Scientific Works) vysshykh uchebnykh zavedeniy, Lithuanian SSR, Stroitel'stvo i arkhitektura, 1968, **7**, 55-61 (Summary of literature).
275. Shil'krot, G.S. The lake as an aqueous natural complex, Izvestiya (Bulletin) of the Academy of Sciences of the USSR, Seriya Geografiya, 1970, No. 2.
276. Shkurkov, V.S. Map for appraising the natural conditions for a human population. Author's abstract of candidate's dissertation, Moscow, 1969.
277. Shubayev, L.P. Symmetry, dissymmetry, and antisymmetry in the geographic mantle. In: Sixth Conference on problems of planetology (VI Soveshcheniye po problemam planetologii), Leningrad, 1968, pp. 33-36.
278. Shubayev, L.P. Symmetry and dissymmetry in the geographic mantle, Izvestiya (Bulletin) VGO, 1970, **102**, No. 2, pp. 107-113.
279. Shuleykin, V.V. Interaction of elements in the ocean-atmosphere-continent system. In: Materials of the Fifth Conference of the Geographical Society of the USSR (Materialy V s"ezda Geograficheskogo Obshchestva SSSR), Leningrad, 1970.
280. Shul'gin, A.M. Introductory lecture to a course on the physical-geographic bases of land reclamation, Vestnik (Journal) of Moscow State University, Geografiya, 1968, No. 1, 10-17.
281. Shul'man, N.K. Anthropogenic landscapes, Zapiski (Notes) Amurovskogo oblasti Muzeya kraevedeniya, 1970, **6**, No. 1.
282. Shcherbakov, Yu.A. Influence of exposure on landscapes. Author's abstract of doctoral dissertation, Moscow, p. 197.
283. Shchukin, I.S. Dessicating influence of deserts in temperate zones on adjacent mountainous areas, Vestnik (Journal) of Moscow State University, Geografiya, 1968, No. 5, 99-101.
284. Encyclopedic dictionary of geographical terms (Entsiklopedicheskiy slovar' geograficheskikh terminov), Moscow, 1968.
285. Yurenkov, G.I. Border of the Valdai glaciation as an important topographical boundary. In: Materials of the Fifth Conference of the Geographical Society of the USSR (Materialy V s"ezda Geograficheskogo Obshchestva), Recorded presentations of a paper by N.A. Gvozdetskiy, K.I. Gerenchuk, A.G. Isachenko, and V.S. Preobrazhenskiy: Current status and problems of physical geography, Leningrad, 1970.

286. Armand, A.D. Natural complexes as self-regulating information systems, Soviet Geogr. Rev. and Transl., 1969, **10**, No. 1, 1-13.
287. Armand, D.L. Studies of the physical processes in the landscape, 1st Internal. Geogr. Congr., India, 1968, Abstracts of papers, Calcutta, 1968, 126.
288. Bertrand, G. The landscape and global physical geography. A methodological investigation, Rev. geogr. Pyrenes et S. Ouest, 1968, **39**, No. 3, pp. 249-272.
289. Boychouk (Boychuk), V.V., Marchenko, A.S. Basis and variations of the physical geographic environment, 21st Internat. Geogr. Congr., India, 1968, Abstracts of papers, Calcutta, 1968, 128.
290. Clark, S.B.K. Landscape survey and analysis on a national basis, Planning Outlook, 1968, No. 4, 15-29.
291. Devdariani, A.S., Greysukh, V.L. The role of cybernetic methods in the study and transformation of natural complexes, Soviet. Geogr. Rev. and Transl., 1969, **10**, No. 1, 14-23.
292. Fines, K.D. Landscape evaluation: a research project in East Sussex, Reg. Stud., 1968, **2**, No. 1, 41-55.
293. Haase, G. Bemerkungen zur Methodik einer grossmasstäbigen landwirtschaftlichen Standartkartierung auf der Grundlage landschaftsökologischer Erkundungen, Wass. Z. Martin-Luther-Univ. Halle-Wittenberg. Math-naturwiss. Reihe, 1967, **16**, No. 5, 669-688.
294. Haase, G. Inhalt and Methodik ein umfassenden landwirtschaftlichen Standartkartierung auf der Grundlage landschaftsökologischer Erkundung, Wiss. Veröff. Dtsch. Inst. Länderkunde, 1968, No. 25-26, 309-349.
295. Handbuch für Landschaftspflege und Naturschutz. Schutz, Pflege und Entwicklung unserer Wirtschafts- und Erholungslandschaften auf ökologischer Grundlage. Buchwald Konrad. Engelhardt Wolfgang. München-Basel-Wien, Bayerischer Landwirschaftsverl. Bd. 2, Grundlagen, 1968, **XIV**, 245S.
296. Jacsman, J. Einführung in die Landschaftsplanung. Zurich. Inst. Orts-, Regional- und Landesplanung ETH, 1966, 63S 10.30 fr Ref. Mitt. Landschaftspflege, 1968, **8**, No. 4, 117.
297. Kalesnik, S.V. The development of general earth science in the USSR during the Soviet period, Soviet Geogr. Rev. and Transl., 1968, **9**, No. 5, 393-402.
298. Kämpfer, M. Landschaftsplanung (deutschsprachiges Schrifttum). Bad Godesberg, Bundesanst. Vegetationskunde, Naturschutz und Landschaftspflege, 1969, **37**, Bl. 4 DM Ref. Mitt. Landschaftspflege, 1969, **9**, No. 3, 95.
299. Kavrichvili, Ketevan. Systemization of mountain regions, 21st Internat. Geogr. Congr. India, 1968, Abstracts of papers, Calcutta, 1968, 131.
300. Kongo, A. Faatsieste iseloomustamisest, Tartu Ulikooli toimetised, 1969, Issue 227, 45-65.
301. Milovidova, N.V. The use of the tree of logical possibilities in the construction and control of physical geographic classifications, Soviet. Geogr. Rev. and Transl., 1970, No. 14, 256-262.
302. Petrov, M.P. Classification of deserts, 21st Internat. Geogr. Congr., India, 1968, Abstracts of papers, Calcutta, 1968, p. 134.

303. Neef, E. Die theoretischen Grundlagen der Landschaftslehre. Gotha-Leipzig. Verl. VEB Hermann Haack, 1967a, 152S, Ill.
304. Neef, E. Die technische Revolution und die Aufgabe der physischen Geographie, Geographie und techn. Revolut., Gotha-Leipzig, 1967b, pp. 28-41.
305. Neef, E. Die geosphärische Dimension in der regional geographischen Arbeit. Przegl. geogr., 1968, **40**, No. 4, 733-746.
306. Palierne, J-M. The concept of the landscape in physical geography: Is it a false problem? Norois, 1969, **16**, No. 62, 254-262.
307. Pauliukevicius, G. Kai kurios renaturalizuoto misko landsafto geochemines savybes, LietTSR Geogr. draugija. Geogr. metrastis, 1967, **8**, 19-26.
308. Richter, H. Naturräumliche Strukturmodelle, Petermanus geogr. Mitt., 1968a, **112**, No. 1, 9-14.
309. Richter, H. General and special structural models of the homogeneous natural area, 21st Internat. Geogr. Congr., India, 1968, Abstracts of papers, Calcutta, 1968b, 135-136.
310. Schmidt, G. Die Entwicklung der physischen Geographie in der Sowjetunion im Spiegel ihrer Publikationen, Geogr. Ber., 1967, **12**, No. 45, 322-337.
311. Schmithüsen, J. Begriff und Inhaltbestimmung der Landschaft als Forschungsobjekt vom geographischen und biologischen Standpunkt, Arch. Naturschutz und Landschaftsforsch, 1968, **8**, No. 2, 101-112.
312. Troll, C. Landschaftsökologie und Biogeocenologie. Eine termenologische Studie, Rev. roum. geol., geophys. et geogr. Ser. geogr., 1970, **14**, No. 1.
313. Varep, E. Maastikuteaduse arengusuundadest eestis, Tartu Ulikooli toimetised, 1969, **227**, 3-12.
314. Vinogradov, B.V. The geographical zonality of the African sector of the Earth according to global space photography from the spacecraft Zond-5, 1968.
315. Wormbs, R. Landschaft als ökologisches System, ARCH, 1969, **2**, No. 6, 16-18.

ECONOMIC GEOGRAPHY
(A Survey of Basic Tendencies, 1966-1970)

A. A. Mintz

The current stage of the development of Communism in the USSR and the accumulation of knowledge deriving from the scientific and technological revolution determine the general conditions of, and create a favorable environment for, the development of Soviet economic geography. Processes of social evolution have caused economic-geographic research to increase considerably in importance. The significance for Soviet society of the problems studied by this science is convincingly demonstrated by the record of the 24th Congress of the Communist Party of the USSR.

The profound changes resulting from the influence of scientific-technologic and social progress in the very subject matter of economic geography pose for this science the task of diligently studying these changes. At issue here are the important and sometimes fundamental changes occurring in the territorial organization of the material and technological basis of production, the profound shifts in the material and spiritual requirements of human beings and the changes in the mode of life resulting from them. Control and planning of these processes cannot be successful without detailed analysis, appraisal, and prediction.

New problems in the sphere of objective reality put greater demands on economic-geographic knowledge and methods of research. The complex character of a number of these problems stimulates integrational processes in science. As a result, "intermediary" and multiple disciplines possessing great "integration potential" appear, as it were, in the center of a developing group of mutually cooperating sciences. Economic geography may be regarded as a discipline of this type. An example of a similar situation is the investigation of the problems, or, more precisely, the complex of problems, involved in the interaction of society and the natural environment, which has now taken on great theoretical and practical importance. An analogous state of affairs obtains in the study of problems of urbanization.

The importance of economic-geographic research directed toward new solutions to traditional problems (analysis and prediction of the territorial structure of production), and also toward the formulation and solution of new problems (the territorial aspects of the interaction of society and the environment, the development of services and branch facilities for the satisfaction of new social requirements, etc.), is placing an increasing "social demand" on economic geography. This demand finds its concrete expression in the singularly broad scope of territorial planning, which requires an entirely different level of scientific support than has been the case in the past.

From another point of view the general development of science in our time has revolutionized the resources available to researchers. Both general scientific techniques such as the systems approach, cybernetics, information theory, mathematical logic, etc., and the fundamental methodological revolution, the development of instrumentation, technology, and investigation procedures based on mathematical-statistics analysis and mathematical modeling are to be mentioned in this regard. This methodological revolution is founded on a fundamentally new material basis: computer technology. And, although new features are expressed in different ways in different branches of science, no discipline, including economic geography, can remain unaffected by this process.

The above discussion permits us to draw the conclusion that the joint influence on the progress of economic geography of external factors and processes of the internal development of science, as reflected in Soviet literature, has created the conditions for a major increase in the social significance, scale, and qualitative level of economic-geographic research.

At the same time, however, a discrepancy in the actual position of economic geography relative to the available objective prerequisites of its development appeared with unmistakable clarity during the period in question. The primary issue here is the official, i.e., organizationally determined, economic geography, represented by a relatively small number of geographic institutions of the academic type (the Geographical Institute of the Academy of Sciences of the USSR, the Geographical Institute of Siberia and the Far East of the Siberian Section of the Academy of Sciences of the USSR, the Geographical Institute of the Academy of Sciences of the Georgian SSR, of the Academy of Sciences of the Azerbaijan SSR, etc.) and a larger number of departments of geography in universities, pedagogical institutes, and other institutions of higher education.

Socially necessary research of a basically economic-geographic character developed to an extraordinary degree outside official economic geography, primarily within the framework of the system of economic institutions. The most powerful, though not the only, organizational basis for this work was research undertaken in accordance with the decision of the party and government on the creation of the General Plan for Allocation of the Productive Resources of the USSR. This research, covering a wide range of interrelated branch, regional, and multiregional problems, was directed in its scientific-methodological and organizational aspects by the Council for the Study of Productive Resources (SOPS) of Gosplan (State Planning Commission) of the USSR and the corresponding Scientific Council in the Economic Division of the Academy of Sciences of the USSR [148, 150, 151]. During the period in question, research conducted both in accordance with the General Plan and independently of it, interest on the part of economists in the territorial, including the regional, aspects and problems of social production, grew considerably.

However, this broadening of the economic-geographic research front was accompanied by an overt and covert denial of its relevance to economic geography as a science. In the past economists, in dealing with the problems involved in the allocation of production within the framework of economic geography, held that this discipline was one of the economic sciences. Economic geography is now no longer mentioned, at least in publications with an economic format.* In addition to the term "allocation of productive resources," which economic geographers had always frequently used to designate a specific scientific discipline and the activities corresponding to it, the terms "regional research" and "regional economics" began to be used [151, 167]. As a rule, the question of how to distinguish the content of these disciplines from that of economic geography has received little attention, although attempts have been made to draw a demarcation between them [166].

The question of distinguishing regional economics from economic geography would not be of great significance if it were a matter only of the term used to designate a scientific discipline. Unfortunately, whatever the objective and subjective causes of this split may be, great methodological experience and valuable concrete results of economic-geographic research are being ignored as a result of this approach, which could bring about a distinct loss for both theory and practice of the allocation of productive resources.

On the other hand, in the geographical literature authors sometimes attempt to ignore these tendencies and automatically classify all pure and applied research in the field of the allocation of productive resources as part of economic geography [97].

This process of misnaming economic-geographic research is also fed by the relative weakness of economic geography in the system of the geographical sciences, by its organizational separation from theoretical and applied activity of an economic character, and by the separation between university-affiliated economic geographers and those affiliated with the Academies of Sciences. The dominant position which the natural scientific disciplines have

---

*It should be noted that a complete break has not occurred here. In participating in geographical forums, both Soviet and international, and also in their publications of a geographical character, leading allocation economists indirectly assert the connection between their activity and geography, and, consequently, economic geography.

traditionally enjoyed in Soviet geography and its consequent inclusion in the organizational system of the natural sciences has undoubtedly considerably influenced the development of economic geography and its place in science.*

These conditions lend particular urgency to the analysis and evaluation of basic tendencies in the development of the theory and methods of economic geography. It is appropriate for this purpose to take into account work which does not have the external characteristics of economic geography, but which in essence belongs to its historical sphere of activity. Such analysis is facilitated by the fact that, in the past few years, fundamental survey research has been carried out on both general and specific problems in Soviet economic geography [138, 172, 173, 185, 201].

At the same time, the breadth of economic geography as a set of interrelated disciplines, the presence in it of various viewpoints with regard to many cardinal methodological questions, and, finally, the scope of its literature significantly hamper the accomplishment of this task. For this reason the present survey cannot pretend to provide an exhaustive coverage of the field, and the selection, interpretation, and evaluation of sources is understandably not entirely free of the influence of the subjective views of the author.

GENERAL QUESTIONS IN THE THEORY AND METHODS OF ECONOMIC GEOGRAPHY

Although in recent years discussion of the definition of the subject matter of economic geography and its relation to other sciences, especially adjacent disciplines, has become less sharp, it nevertheless continues in the literature.

The celebration of Lenin's 100th birthday stimulated great interest in the analysis of the influence of Lenin's scientific legacy on the development of all the sciences. This legacy was especially significant for Soviet economic geography, a young science whose genesis and development proceeded under the direct influence of Leninist economic and philosophical theory, as well as Lenin's plans for the creation of a Communist society in our country.

The various forms of Lenin's influence on the development of economic geography were analyzed in publications written in honor of this event, which also attempted to interpret contemporary achievements and problems in this field in the light of Leninist theory [4, 10, 26, 71, 74, 96, 168, 176, 188, 217, 230, 236, etc.].

It should be noted, however, that we do not as yet have any major monographs on the theory of economic geography based on Marxist–Leninist philosphy, critical analysis of the development of this science in the

---

*In this respect Soviet geography is sharply distinguished from most foreign schools, in which economic geography (in the broad sense of the term) either wholly (as in the US) or partly dominates.

Soviet Union and abroad, the achievements of adjacent disciplines, the usefulness of general methodological conceptions, the development of research techniques, or experience in the practical application of economic-geographic knowledge.

Theoretical and methodological efforts in economic geography are still concentrated on specific, although undeniably important, problems.

The *definition of economic geography* adopted by the Second Conference of the Geographical Society of the USSR retains its significance: "Economic geography is social science, forming a part of the complex of the geographical sciences, and having as its object of study the geographical allocation of production, the latter understood as the totality of productive resources and productive relations, and the conditions and characteristics of their development in various countries and regions" [247, p. 423].

At the same time, another view of the subject matter of economic geography continues to exist, based on two fundamental propostions: a) the inherent relevance of this discipline to geographical science, at the center of which should be the territorial aspects of the interaction of nature and society; and b) the need for the existence of a specific material object of study. According to this view, then, the objects of economic-geographic investigation are "economic systems with actual spatial existence (including productive-territorial complexes) of various scales and types" [244, p. 12].

The author of this definition, Saushkin, later noted that less disagreement occurs regarding the concrete content of the tasks facing economic geography as a science than regarding the concise definition of this content. He argued against the "allocation" definition, asserting that the allocation of production is a practical task for a number of sciences, and defined economic geography as the "the science that studies complex dynamic territorial systems of productive resources in relation to their environments – both social and geographical" [207, p. 42].

Earlier Kolotiyevskiy [90] formulated the position that the general theory of allocation of production (which is a complex economic problem requiring the efforts of many sciences for its solution) should be studied by economic geography in the broad sense of the term, while economic geography in the narrow sense should deal with its particular subject matter, the productive-territorial systems (complexes) that develop as a result of territorial differentiation and coordination of labor made possible by increasing mastery of the geographical environment.

Thus, there exists the possibility of achieving unanimity not only in the area of concrete research, but also with regard to definitions that have been debated for decades. The difficulty in this regard, however, consists in the fact that, as was noted above, work on the general theoretical problems of allocation of production and investigation of territorial formations – plexus points, complexes, regions – is being done today in the framework of the economic sciences, in which disciplines are developing that duplicate the work of economic geography. The resulting debate concerning the possibility of defining the boundaries between allocation economics and geographical economics, the latter concentrating on territorial complexes, has not been justified in practice over the last few years.

The significant difference between Soviet approaches in any of their variants, from those prevailing in bourgeois geography, which takes as the geographical criterion primarily the spatial approach to things, i.e., the chronological aspects of research, rather than any actually existing material object, should be noted. The expansion in this instance of the objective basis of the spatial approach is resulting in an expansion of foreign geography as a social science. Economic geography in the USSR covers a much narrower range of problems, both in theory and in practice, than in the West. Not all spatial forms of human activity by far are becoming the object of economic-geographic study.

At the same time, basically as a result of social processes and practical requirements, the interests of economic geography are going beyond the phenomena of material production to the broad range of social phenomena, as well as to interaction with the natural environment. Thus, economic-geographic research is coming to include the entire process of social productivity in the broad sense of the term.* This process gives rise to new trends in economic geography, to the strengthening of links not only with traditional neighbors — economic and physical geography — but with other social disciplines as well — sociology, demography, etc. [2, 178]. This standpoint raises the question of the correctness in preserving the understanding that has developed historically in the Soviet Union, namely of economic geography in the broad sense as the complex of all the social geographical sciences. There has not occurred as yet, however, any extensive discussion of this question and of proposals for a new system of classifying social-geographic research (corresponding to human geography abroad).

Disagreements regarding the relation of economic geography to the other sciences have noticeably lessened. With the exception of individual shades of opinion, there is general agreement that economic geography, by virtue of the basic nature of its subject matter, is a social and economic science that also enters into the system (complex) of the geographical sciences. This view is, to a large degree, the result of increasing understanding of the integrational tendencies that are developing in science in general, and that reflect the growing importance of complex interdisciplinary research into the most complex and acute problems of the current age.

On the other hand, it is has become clear that the idea of a unified geography as a separate discipline oriented toward creation of integral

---

*In part this broadening of the research front is encouraged by the desire to investigate aspects of society that have been little studied from the geographical point of view, and thereby to avoid the increasing competition with economists in the traditional areas — the study of economic regions and specific branches of industry.

geographical characterizations has become anachronistic.* Even in the period under consideration here, however, there have appeared echoes of the former dispute which preserved the basic features of the earlier debate [13, 216, etc.].

The ideas of constructive geography, including formulation of major problems in the use of natural resources and the transformation of nature, productive–territorial complexes, and settlement patterns [34] formed the basis for the development of qualitatively new links between natural science and economic geography within the framework of the system of the geographical sciences. The extreme antigeographical assertions that set physical geography off against economic geography and denied any substantive link between them have almost disappeared from the literature.

The question of the relation of economic geography to the other sciences received rather detailed treatment in Kolotiyevskiy's monograph with regard to the general theoretical problems in this science. Full accord has not yet been achieved, however, with regard to the degree of internal unity and the relation between the processes of integration and differentiation in the system of the geographical sciences. Thus, in an article published in Hungary and dealing with Soviet area studies, Anuchin described this issue as the most centripetal in geography [249]. Rodoman [196] believes that the existence of the anthroposphere, consisting of the geographical mantle together with human society and the results of its activity, requires a separate complex geography in addition to physical and economic geography.

Semevskiy has considered the question of the structure of geography, and has sharply criticized unified geography as a study incompatible with Marxism–Leninism, since it assumes the extension of natural laws to human society. Its integrity as a complex of sciences should, he claims, be understood dialectically, as a unity of opposites, namely two systems of sciences – physical–geographic sciences and economic–geographic sciences – combining in a new way the principles inherent in each [214, 216].

Many of the general problems of economic geography were debated with renewed vigor in connection with the *influence of new general-methodological approaches,* as well as *new methods of research, on its content.* The specific aspects of this question are associated with the influence on Soviet science of the significant progress occurring during the 1960s in foreign, especially American, geography, and which was bombastically designated as a quantitative revolution. The appearance of translations of several works that expressed quite clearly the views of geographers in the capitalist countries (or rather, the most influential trends in their thinking), such as Bunge and Hagget, and of a collection of articles entitled *Models in Geography,*

---

*In spite of the intense polemics of the late 1950s and early 1960s, concrete research in the spirit of a "unified geography" demonstrating the advantages of this approach has not appeared. Increased interest in theoretical, methodological, and problem-oriented applied research has overshadowed the entire "characterization" genre.

permitted Soviet geographers to gain acquaintance with the concepts of bourgeois geography.

Insufficient attention to critical analysis, evaluation, and philosophical interpretation of the new trends in contemporary foreign geography characterized the commentaries (forewords and concluding remarks) by Soviet geographers (Gokhman, Medvedkov, and Saushkin) that accompanied these translations. On the other hand, serious critical consideration of foreign methodological concepts, the drawing of distinctions between ideologically unacceptable philosophical positions and useful techniques and approaches to research were at times subordinated to an attempt to reduce new features of scientific theory and methodology to the categories of the previous debate over a unified geography, and application of natural laws to human society, etc. [11].

The introduction of new concepts with regard to the systems approach, structural–functional analysis, cybernetics, information theory, as well as the application of mathematical statistics, mathematical logic, modeling, and other techniques of formalization and quantitative analysis, which have greatly influenced many natural and social sciences, is proceeding in economic geography with great difficulty. On the one hand, there is a tendency to use new methodological approaches that go beyond the limits of economic geography per se and which are proclaimed as the basis of new, general-, or supra-geographical disciplines, such as mathematical geography, theoretical geography, and metageography [40, 47, 209, 210]. It has been suggested that the broad application of mathematical tools is opening the way towards the creation of a specialized theory within the scope of economic geography [132, 133]. On the other hand, a reaction to these claims is appearing in the form of a tendency to deny or belittle the influence of new conceptions on the methodology of the geographical sciences, to reduce their role to that of the improvement of methods and research techniques [12], to treat certain claims as contradictory to dialectical materialist philosophical principles, and to make implicit inferences regarding the solubility of fundamental methodological problems in the geographical sciences [76, 77, 109].

In spite of the fact that the central conceptions of Soviet economic geography (primarily the concept of the region as a production–territorial complex and an element in a unified national economy) are extremely close to the concepts of the systems approach, the introduction of the latter into the methodology of economic geography, in spite of the positive experience of other sciences in this area, has not yet begun. There has occurred some interpretation of individual phenomena from the viewpoint of the systems approach (which will be discussed below) and a wide adoption of systems terminology.*

---

*Another aspect of this question should be noted. Theoreticians of the systems approach, using experience accumulated by the biological, economic, and engineering sciences in applying this approach, do not touch on the interpretation of the geographical study of productive and natural territorial complexes, population distribution systems, etc.

Although considerable material was published on general theoretical questions, there was, as before, relatively little research devoted during the period in question to the development of the conceptual apparatus of economic geography, to its central conceptions. Significant material of this sort was presented in Kolotiyevskiy's monograph [90]. Individual efforts were made to deepen the extremely important concept of the economic–geographic situation to establish its content [122] and enumerate its quantitative indicators [66]. The concept of the area function was used to generalize the theory of statics and dynamics of allocation of various public services [141].

Among the new issues facing economic geography is the extension of the concept of applied geography, which has received wide currency abroad but has been little utilized in the Soviet Union [122],* and the suggestion that economic–geographic research is important in the theory and practice of prediction [60, 61, 203, 205]. Closely associated with the development of theoretical conceptions and the realization of the practical goals of economic geography is the *improvement of its methods.* The most striking feature of the period in question in this regard was the dissemination of *mathematical methods:* the range of theoretical and applied problems in which mathematical methods were used expanded; the methodological arsenal of the field was increased both through the borrowing and through the creation of mathematical tools; the number of investigators working on the development and, especially, the application of mathematical methods increased; increased organizational measures were taken – special teams of researchers were created, lecture courses, conferences, seminars, and "summer schools" were instituted, collections of articles were published, etc. Work was done in the classification of the models used in economic geography [57]. The current state of mathematization of economic geography was examined at the Fifth Congress of the Geographical Society of the USSR [208]. This question is considered in a separate article in the present survey, and need not be examined in detail here.

All things considered, it cannot be claimed that Soviet geography is experiencing the quantitative revolution that, for about 10 years, has been ascribed to the activities of a number of foreign (especially American, Swedish, English, etc.) geographical schools. The central concern in methodological work, therefore, is the improvement of applied traditional methods.

The strengthening of the analytical and problem-oriented applied approaches in all areas requires increased attention not so much to the formal characteristics of descriptions as to the demonstrability, clarity, and specificity of the statements made. For the most part traditional, but formerly insufficiently applied, methods of statistical analysis – tabulation, calculation of averages, indices, correlation functions, etc. – are being used for this purpose.

---

*It has been suggested that this term is unsuitable because it tends to set theory against praxis [91].

The cartographical method, specific to the geographical sciences, also continues to be significant. Two basic tendencies have characterized economic cartography in the past few years.

In the first place, in accordance with the differentiation of economic-geographic research, the subject matter of maps is rapidly expanding — new categories (for example, maps providing economic appraisal of natural conditions and resources), as well as new cartographical subjects in the context of developing areas of research (agricultural maps, etc.), are appearing. This tendency has manifested itself quite clearly in cartographical publications, especially complex regional atlases; it is reflected also in the rather numerous conferences on cartography and in articles by geographers and cartographers [78, 135, 163, 169, 221, 226, etc.].

Second, attempts are being made, in close association with the above-mentioned expansion of cartographical research and content of economic maps, to improve methods of representation, i.e., to continue the work begun by Baranskiy. Although classical methods of representing economic phenomena predominate in most cartographical publications, some rather interesting innovations have appeared [154].

Relatively recently there appeared a substantive analysis of the current state of the most important issues in economic cartography [125], and so we may limit the present discussion to the general considerations presented above.

In light of new methodological approaches the old methods are acquiring a new significance: maps and graphical representations of structure and interrelations that have long been used in economic geography are being utilized as a type of model [48, 57, 75, etc.].

As was noted above, the differentiation of economic geography, the separation and development as separate disciplines of its different trends, the number of which is already significant [90, 201], is proceeding slowly. In respect both of organization and personnel these trends are already quite interwoven, so that in scientific institutions and institutions of higher education an undifferentiated economic geography predominates.*

Nevertheless, in the interests of analysis we may distinguish several general trends in economic-geographic research.

ECONOMIC REGIONALIZATION AND THE STUDY OF PRODUCTIVE-TERRITORIAL COMPLEXES

In spite of the appearance of new objects for investigation and corresponding new research trends, investigation of the territorial structure of the economy, primarily its regional aspect, continues to be central among

---

*A well-known exception is the Institute of Geography of Siberia and the Far East of the Siberian Division of the Academy of Sciences of the USSR, which contains as part of its structure several specialized economic-geographic units.

the problems with which economic geography deals. Clearly, therefore, it is precisely in this area that all of the tendencies characteristic of the current stage of development of Soviet economic geography appear especially clearly.

Topics of research have begun to show a clear shift away from economic regionalization in the narrow sense of the term, i.e., consideration of the concrete network of economic regions, to the broad range of problems involved in the study of territorial economic organization, of the economic structures on which the regions are based, i.e., productive-territorial complexes of various orders and types.

In regard to *economic regionalization,* a certain accord has been reached concerning its basic methodological principles as formulated by Pokshishevskiy [172] on the basis of a generalization of material emerging from the lively debates of the late 1950s and early 1960s.

Kolotiyevskiy's generalization of the scientific and practical experience of Soviet economic regionalization [90] likewise did not evoke serious objections. This author's contribution consisted especially in his clarification and development of the conceptual basis of the theory of economic regionalization, which is of no small significance considering the high degree of subjectivism characterizing the understanding and application of certain concepts and terms in this area. Thus, the interesting proposal was made to distinguish regionalization as an objective process, regionology as the designation of the corresponding branch of scientific investigation, and, finally, regionalization as a societal-practical activity. Kolotiyevskiy's discussions and proposals regarding the methodology of regionalization, the classification of economic regions, etc., are of great interest.

Vetrov [27] approached the methodology of economic regionalization from the viewpoint of historical analysis of the evolution of fundamental scientific ideas and concepts. Although this author's approach is based on analysis of a considerable range of literature, especially economic-geographic literature, his conclusion that economic regionalization is a separate science dealing with the territorial division of labor and not coinciding, in spite of their mutual subject matter and methods, with economic geography, can hardly be said to be consistent with the overall development of Soviet economic geography.

Unfortunately, Kolotiyevskiy's and Vetrov's monographs, which were published in small editions (1000 copies) by local publishers, did not receive the attention they deserved, given the breadth of the questions they formulated. It is clear that there has also been a certain decrease in interest in economic regionalization due to the transition to the branch system of industrial management, the stabilization of the network of economic and administrative regions, and a number of other factors.

Nevertheless, it would be incorrect to conclude that methodological efforts directed specifically toward economic regionalization have ceased altogether. Thus, methodological works by leading workers in the field of economic regionalization have appeared [88, 238]; the taxonomy of integral economic regionalization [100], the content [158] and methodology [79] of fractional regionalization [108], and the links between fractional economic regionalization and regional planning [114] have been discussed.

A series of lectures on questions of economic regionalization, based on recent theory and practice [73] as interpreted by Kolosovskiy at Moscow State University, has appeared. In these lectures the opinion was advanced, based on historico-geographic material, that it is incorrect to consider the territorial division of labor as the only regionalizing factor and that the appearance of economic regions is associated with the development of capitalistic economic relationships [72]. This same author has also surveyed the state of economic regionalization in countries with various socio-economic systems in the light of Lenin's methodological legacy [74]. Regionalization nuclei or centers were discussed and a methodology of regionalization based on the use of quantitative methods was proposed [160].

In developing his previously expressed views, Rodoman [195] deepened his conception of the formal-logic aspect of regionalization, and applied the tools of set theory and graph theory to the solution of these problems. He also advanced the concept of socio-geographic regions, their classification (functional, complex, plexal) and the resulting concept of hierarchies of regions and territorial organizations as systems of functional-plexal regions [197]. Although a number of the positions expressed in these works contradict methodological positions previously formulated by certain theoreticians of Soviet economic regionalization, the absence of the conditions for open debate makes it impossible to evaluate Rodoman's arguments. Of undoubted interest is the considerable attention this author devotes to the morphological aspect of the processes in question.

Attempts were made to apply mathematical tools to the formal side of the regionalization process — the division into territories as a function of specific features. In this connection utilization of the techniques of division into groups of finite point sets, reducing the process to a linear integer programming model, and application of an information-theoretic approach to the solution of the resulting problems were proposed [155, 156].

Much more intensive research was carried out during the period under consideration on *problems of the territorial structure (organization) of the natural economy,* especially on various problems having to do with productive-territorial complexes (PTCs).*

---

*There is no need here to consider this group of problems in detail, inasmuch as they have been fully dealt with in a recently published survey covering a wide range of the literature (959 titles), mostly appearing during the period under consideration (1965-1969).

In connection with the growth of scientific and practical interest in the development of PTCs, the Council on the Study of the Productive Resources of the Ukrainian SSR of the Academy of Sciences of the USSR held a special conference on these questions in 1968. Among the authors of the resulting collection of articles, which dealt with the current state of study of PTCs, their nature, and their classification, were Palamarchuk, Silayev, Shrag, Maslov, Lasis, and others [59].

The study of PTCs, which are the most effective organizational form for planning the territorial organization of productive resources, occupies a central position in research on this problem. This study goes back to the ideas of Lenin, to the experience, theory, and practice of Soviet national economic planning; its fundamental theoretical and methodological positions were worked out and formulated by Kolosovskiy, reinforced by Baranskiy, and became part of the foundation of Soviet economic geography. The development and broadening of the PTC concept has been facilitated in recent years by the posthumous publication of a number of works from the scientific archives of Kolosovskiy [88, 89].

The opinion has been expressed that the development of economic-geographic modeling as a specific method for investigating PTCs has three interrelated aspects: modeling of territorial resource sets, carried out primarily in the geographical institutions of the Academy of Sciences of the USSR, modeling of branch productive enterprises in the PTC system, and modeling of entire systems of complexes, carried out in universities and scientific centers associated with national economic planning [49].

It has already been noted that the research done by many Soviet economic geographers, especially those associated with the Department of Economic Geography at Moscow State University and the Institute of Complex Transport Problems of Gosplan USSR,* covers many individual aspects of the PTC concept, but devotes insufficient attention to the most general questions having to do with the concept as a whole and the philosophy of its general methodology [80]. The attempt to consider fundamental assumptions of the PTC concept in light of the methodology of the systems approach, as a complex system to whose study the appropriate methodological apparatus may be applied, is of considerable interest in this regard [64, 74, 75, 75a]. This approach is the more justified in that the PTC concept and a number of its categories are, in essence, very close to general methodological ideas and concepts of modern systems theory. It is this methodological affinity that has permitted the modeling method to be fruitfully introduced into the study of PTCs, including its most complex mathematical variants [48, 56, etc.].

---

*Two issues of *Voprosy geografii* (Problems of Geography) were devoted to these questions, with the participation of members of these organizations: No. 75, The Territorial Organization of the Productive Resources of the USSR, Moscow, 1968, and No. 80, Territorial Productive Complexes, Moscow, 1970.

A methodologically important, if not major, part of the study of PTCs — Kolosovskiy's *concept of energy-production cycles* — has developed within the scope of the systems approach. Its genesis and development in connection with the study of PTCs, as well as criticism by its opponents, have received comprehensive treatment by Stepanov [223]. The key role of this concept in the study of PTCs and its great methodological significance in the investigation, modeling, and construction of complexes has induced economic geographers to devote their attention to its elaboration. Both the profound changes in the technology of production (and the role of the technological factor in the concept of cycles can hardly be overestimated) and the unequal elaboration of the various generalized cycles in Kolosovskiy's publications serve as the basis for this interest. Unlike critics of this concept who have used the recent situation only to argue against it, Kolosovskiy's followers are attempting to preserve and develop all of the strong features of the idea of energy-production cycles.

Perhaps the most significant aspect of the activity of this school of economic geography is its use of the method of cycles in the analysis and modeling of PTCs [206, etc.] as an alternative to the traditional and widely known analysis of the structure of branch production facilities. Efforts at qualitative characterization of regional PTCs of various types were made, based on the concept of the energy-production cycle [113].

There have also been attempts to modernize the cataloging of cycles. Since Kolosovskiy was concerned mainly with industrial-type cycles, attempts were made to develop his ideas as applied to agriculture [63]. Two approaches to the cataloging of concrete cycles may be noted here. The first consists in retaining the system of generalized cycles and their aggregates in the form in which Kolosovskiy proposed them, isolating the smaller cycles and branches [75]. The second approach consists in broadening the catalog of cycles — as Stepanov [224] expressed it, in breaking down or concretizing them [202, 204, 235]. It is true that the principle of the unity of the directing of energy does not always hold, but subdivided cycles are not far from ordinary branches of production. Efforts were made to combine branch and cyclic analysis of production complexes [153, 241a]. The proposal was made to distinguish different stages in the development of PTCs, from simple territorial industrial groupings to comprehensively developed complexes [239], as well as various types of interaction between enterprises within a PTC [248]. New developments of the concept of the energy-production cycle concept were proposed and elaborated [224a].

Another direction taken by research, and one which has attracted considerable attention on the part of economic geographers, is the *study of minor territorial links of regional* PTCs, primarily industrial blocks. The literature dealing with the study of *industrial blocks* is quite extensive, but definitions of what constitutes an industrial block in economic–geographic works do not by any means always coincide with one another [7, 37a, 224, 235, 237, etc.].

Interest in territorial complexes during the period under consideration was apparent in research carried out by allocation economists, especially in the publications of the Council for the Organization of Productive Resources of Gosplan USSR and the Institute of Economics of the Academy of Sciences of the USSR. The major reason for this is no doubt the objective importance of the problem — the development of PTCs as the territorial organization of productive resources at the current stage of the building of Communism in the USSR. Recognition of this situation has resulted in the development of research on territorial complexes, the more so that practical requirements must be met — the necessity of developing the regional sections of the General Scheme for the Allocation of the Productive Resources of the USSR [8, 58, 81, 183, 220, 241 et al.].

The question as to whether research into PTCs carried out by economists differs from that done by economic geographers is of interest from the methodological point of view. Unfortunately, the differing views of the representatives of each of these two groups make it impossible to arrive at a precise answer to this question. Moreover, imprecise definitions and loose usage of terms, especially terms such as "complex," give rise to considerable confusion. Stepanov [224], after considering these questions and noting the variation in the understanding of these terms, concluded that Soviet geographers are in complete agreement regarding the basic content of the concept of the complex.

A somewhat different linear chronologic solution to this problem has been proposed by Shrag [241], who noted the pioneering role of Kolosovskiy in the theoretical investigation of PTCs; he wrote that "economic geographers subsequently made significant contributions through their study of the economic development of individual regions of the USSR, but did little to further develop the theoretical positions advanced by Kolosovskiy. Economists made a significantly greater contribution to the solution of problems in the complexing of industrial development during the postwar years" (page 9). This description hardly indicates an adequate knowledge of the current state of economic–geographic research.

According to this view, economists, in the course of developing the theoretical and practical issues of allocation of productive resources, arrived by means of pure logic at the idea of the territorial complex as an objectively existing structural element in the national economy, and also at the importance of complexing all specialized components of the economic structure. To a much greater degree than economic geographers, they have devoted and continue to devote attention to the economic efficiency of the territorial organization of production. For this reason alone the most far-reaching analysis was aimed at those complexual aspects of the economy that permit a determination of economic capacity. The necessity of utilizing available information for this purpose (although traditional sources must also be taken into account) caused branch analysis to predominate over the method of energy-production cycles. It is clear that the same thing occurred with regard to the analysis of industrial blocks, which was primarily based on the assumption of the advantages of integrated construction and maintenance of groups of enterprises. This definition of the industrial block came to predominate in the literature dealing with planning.

It should be noted that these considerations are most applicable to analysis of developing economic regions, in which the complex is, as it were, identical with the economy of the region. With regard to newly forming PTCs, however — the Bratsk, Sayansk, South Tadjik, etc., PTCs — the difference between the approaches decreases. At this point it becomes possible to utilize all of Kolosovskiy's basic concepts. The retention of previously formed demarcation lines and the rejection of foreign ideas in this area can only diminish effectiveness in practice.

The *mathematical economic* approach is a particularly important one. It arose within the scope of research in modeling the development and allocation of productive resources performed by the Central Institute for Mathematical Economics of the Academy of Sciences of the USSR and in the Institute for the Economics and Organization of Industrial Production of the Siberian Division of the Academy of Sciences of the USSR in Novosibirsk. These researchers go back to the first experiments in mathematical economic modeling that were performed in the late 1950s and early 1960s under the direction of Nemchinov.

One of the lines of research followed was the utilization of interbranch equilibria at economic regions as models reflecting their internal and external structure [103]. In this approach the territorial scope of the region is assumed given, which makes it inapplicable to the solution of problems in economic regionalization. Another approach links the modeling of territorial complexes with optimization models for the allocation of productive resources. Here the ideas of the economic geography school find wider application. At the present time not only have general methodological formulations of the problem been developed, but considerable experience has also been accumulated in the optimization of concrete PTCs [14, 15, 16, 42, 164, etc.]. It is interesting to note that in publications of the latter type, elements of economic-geographic analysis are retained along with a number of economic approaches, and the authors acknowledge this link with economic geography.

The development of the methodology of research on PTCs was reflected in the content of publications dealing with the study of specific complexes — economic regions, subregions, blocks, and local combinations of production facilities. The methodological tools developed by Soviet economic geography are being applied to an increasing extent in this research. A monograph devoted to the PTCs of the Armenian SSR [24] is noteworthy for its close attention to the methodological aspect of investigations, its innovative approach, and thoroughness.

THE GEOGRAPHY OF BRANCHES OF PRODUCTION

Unlike the issues discussed above, in which until recently research of an economic–geographic nature predominated, research on the allocation of branches of production has long been a field profoundly influenced by the studies of specialists in branch economics, while associated general methodological issues have been dealt with by allocation economists. The compilation of the General Scheme for the Allocation of the Productive Resources of the USSR has reinforced this situation.

On the methodological level it is interesting to note the development of a unified methodology for analyzing the allocation of branches of the economy as carried out by the Council for the Organization of Productive Resources in its capacity as the chief organization dealing with this problem [162]. Major monographs and collections of articles on allocation problems appeared, based on generalization of the enormous amount of information generated by the Soviet national economy, including information accumulated by work on the General Scheme [44, 181, 182, 184, 230, 245 et al.].

Although it is too soon to speak of the creation of a precisely formulated general theory of the allocation of socialist production with a unified system of concepts, categories, and principles, important steps in this direction have been taken [149].

The laws, principles, and factors governing the allocation of productive resources deserve closer attention [230]. The most extensively studied are *allocation factors* — the technical and economic characteristics of branch enterprises, the natural and economic conditions determining allocations for various regions, and scientific and technological progress. Considerable success has been achieved in the quantitative analysis of the mechanism of action of individual factors based on generalization of the extensive material deriving from planning projects.

The great importance of the allocational aspects of the general problem of raising economic efficiency of social production as noted in the decisions taken by the 23rd and 24th Congresses of the Communist Party of the Soviet Union has caused considerable attention to be devoted towards rationalizing the territorial organization of production, and towards developing methods necessary for analyzing it. In this connection, systems of quantitative indicators have been developed that permit various allocation alternatives to be compared [182, 184, etc.]. A critical analysis of foreign theories and methods in the area of allocation of production has also been performed [191].

In the course of solving these problems in preplanning research, which constitutes the basis for territorial planning, the previous weaknesses of allocation research were largely overcome: descriptiveness, the artificial separation of problems involved in the allocation of production from general issues in the development of the national economy, inadequate consideration of the economic efficiency of allocation alternatives, the separation between branch and regional approaches, underestimation of the role of natural factors, etc. A convergence of different directions in research can be observed: problems in the allocation of individual enterprises and branches became more closely associated with general problems having to do with the optimal development and functioning of the entire national economic system and its subsystems — the interbranch and regional blocks.

A basic conception of this convergence began to develop in the form of a *global model for the optimization of allocation of productive forces* of the entire country, as it were, built into a macroeconomic model of the development of the entire national economy. Such a model should include not only newly constructed elements, but should also adequately reflect the long-term functioning of all productive and territorial elements of the national economic system [152]. The development of such conceptions, supported by the required resources and methods and integrating separate analytical investigations (of allocation factors, the territorial forms of branch, interbranch, and regional complexes, technological and economic links, etc.) leads to the development of a broad theory of production allocation. Clearly, such a theory will permit a rigorously formalized description of the fundamental assumptions on which national economic planning is based and the realization of these assumptions in practice.

Worthy of special note in this regard are the successes achieved during the period under consideration in the area of mathematical economic modeling of allocation of productive resources. For a number of reasons problems involved in the allocation of the branches of the economy proved to be the most suitable for the calculations involved in optimization calculations. The solution of these problems in various branches had already begun in the early 1960s. The introduction of mathematical methods and the use of computers in planning proved extremely productive in this class of problems. The Novosibirsk Mathematical Economics Center accumulated and generalized an especially large amount of experience in this area [1, 43, 87, 137, 142, 164 et al.].

Three types of model were developed for the purpose of optimizing allocation: 1) branch, 2) regional, and 3) complex interbranch and interregional models. On the basis of mathematical economic modeling it was possible to integrate various aspects of economic–geographic research that had been the source of considerable methodological difficulties at the qualitative, logical, and verbal descriptive levels.

It was therefore in those aspects of economic geography dealing with the branches of production that the major breakthrough was achieved, albeit primarily through the efforts of economists, and the new methods that quickly resulted found application not only in planning research but in planning practice as well. Clearly, one of the reasons for this development was the adaptability of previously existing mathematical apparatus (linear programming, etc.) to the solution of problems in allocation.

No major methodological progress in research on the branches of production occurred within the scope of economic geography per se, although in a number of papers valuable generalizations of available data were made, including data from neighboring disciplines. A positive feature in this process was the overcoming of previous tendencies to isolate economic–geographic approaches, their "separations from research in economics."

The most significant point in this area was the *appearance of general studies on the geography of the major branches* of the Soviet national economy: industry [235] and agriculture [193], written by Moscow University geographers in the course of an extended period of research and teaching. In addition, these same authors somewhat earlier wrote a textbook on the geography of the branches of the economy [244] that was significantly different from traditional texts in these disciplines.

All of these publications, in addition to presenting large amounts of concrete material relating to the characterization of various branches of production, formulated important methodological questions regarding the geography of branches of production as a separate division of economic geography.

Thus, A. T. Khrushchev, not excluding from industrial geography the research performed by economic organizations (the Institute of Economics of the Academy of Sciences of the USSR, the Gosplan USSR Council for the Organization of Productive Resources, etc.), characterized an economic–geographic approach to the study of industry. The major content of such an approach was specified as the study of principles and characteristics of the territorial organization of industry, as distinct from the concept of allocation. These distinctions lead to a closer integration of the branch and regional aspects of the problem, and to the analysis of productive-territorial combinations of various types and orders as the major economic–geographic method of research. In this regard, in addition to the traditional characterization of individual branches of production, considerable attention was devoted to the modeling (structural and territorial) of various branches of production and to problems in industrial regionalization [235].

Rakitnikov, on the other hand, does not consider it worthwhile to distinguish the terms territorial organization and allocation, considering them synonyms. He does not consider efficient territorial organization of agriculture to be the task of any one branch of science. Rakitnikov defines the specific research directions that do constitute the substance of this discipline, basing his definition on factual data. Economic interpretation (or appraisal) of the natural environment as a factor in territorial differentiation of productive forms of agriculture emerges as the specific goal of the geographical study of this branch of production. The means for carrying out this study include analysis of various types of agriculture, in particular of various kinds of land-utilization appraisal of factors determining agricultural allocation, historico–geographic analysis, and agricultural regionalization [193].

The articles contained in the collection entitled *Voprosy geografii*

(Problems of Geography), No. 75, 1968, devoted to the problems of the territorial organization of productive resources, exemplify attempts at economic–geographic interpretation of problems in the allocation of branches of production [70, 143, 234]. Research was performed on concentric organization of industry based on natural-resource utilization [180].

As a result of the extensive introduction into geographic research of methods and indices of neighboring economic disciplines and the considerable blurring of the boundaries between them, geographers have continued to devote close attention to regional factors (the regionalization of branches of production, their place in a regional economic complex, etc.) and in a number of instances, especially in the study of agriculture, to the territorial differentiation of the natural environment. It is in these areas that economic geographers have achieved their most significant independent methodological results.

THE GEOGRAPHY OF THE NONPRODUCTIVE SPHERE

Although Soviet economic geographers continued to devote the major part of their efforts to various aspects of material production, research into the geographical aspects of the nonproductive sphere has lately developed to a considerable degree.

This phenomenon is timely for a number of reasons. 1) It is theoretically correct in analyzing social production to include within the scope of research the entire process of social regeneration, including the distribution and use of material goods, and the production and use of nonmaterial services. 2) The programmatic characteristic of the building of Communism defines satisfaction of the material and spiritual needs of human beings as the major goal of production, and the creation of the branches of the economy necessary to achieve this goal is now being set as a practical task, as can be seen from the decisions of the 24th Congress of the Communist Party of the Soviet Union. 3) These social aims require that the lag in the development of research into the nonproductive sphere be overcome.
4) The development of research into the nonproductive sphere in cooperation with sociologists, psychologists, and ethnographers permits the solution of new theoretical problems associated with the geographical interpretation of many forms of human activity (motives, evaluations, behavior, etc.) that were formerly scarcely studies at all. The development of research of this type is an example of the increased attention devoted by economic geographers to the broad spectrum of social phenomena and the expansion of economic geography beyond the limits of a single sphere of production, the so-called sociologization of Soviet economic geography [178]. It has been proposed that territorial variations in the conditions of life of the population should be considered as the subject of a separate discipline – social geography [107].

It should be noted that the rapid growth of the nonproductive sphere demands that it be studied in its traditional economic-geographic aspects as well, as a part of the national economy and, consequently, as a consumer of natural and labor resources, and also as an element in the territorial economic complex interacting with external factors and with other elements of this complex.

The first research of this type began within the compass of established research trends. The close link between service branches and the population and the resulting connection between the forms of localization of services and settlement patterns gave rise to the first methodological treatments of this topic within the framework of population geography [84, 175]. Proposals were made for methods of quantitatively appraising service levels for populations of various territorial units [65, 136].

The development of a separate branch of the geography of the nonproductive sphere, *recreational geography*, began somewhat differently. This discipline studies such important service functions as providing the population with opportunities for rest and relaxation. The special role of the natural environment as the most important resource of the recreation industry, the necessity of finding, evaluating, and reserving recreation territories possessing the appropriate characteristics stimulated the development of the first stages of research within the scope of physical geography. It quickly became clear, however, that the solution of the fundamental theoretical and applied problems of this discipline was impossible without invading the social disciplines – sociology, pyschology, economics, and economic geography – and without the closest possible cooperation with the medical-biological and engineering, especially the architectural and planning disciplines. A research group has formed, although not formally organized, consisting primarily of geographers working on the methodological basis of recreation geography and means of applying its results in practice. A considerable number of publications has appeared dealing with theoretical and methodological questions [36, 37, 62, 145, 179, 225, etc.].

The economic-geographic aspects per se of recreation geography, however, and the question of the place of recreational activity in the general national economic system and in individual regional economic complexes have not been adequately developed.

POPULATION GEOGRAPHY

During the period under consideration this branch of economic geography developed somewhat less intensively than during the proceding 5–10 years, during which time interest in populational topics grew explosively and the basic methodological positions of Soviet population geography were formulated (primarily in the works of Davidovich, Kovalev, Konstantinov, Pokshishevskiy). The results issuing from this stormy period were presented at the Second Interdepartmental Conference on Population Geography in 1967. The state of this discipline, including analysis of its theory and methods, was comprehensively discussed in a number of publications [32, 146, 173].*

Although all of the basic trends in population geography continued to develop during the subsequent period, no significant changes occurred in the methodological basis of this discipline, which had its inception in the early 1960s. In addition to numerous publications dealing with specific regional problems, there appeared publications, both by individual authors [50] and by groups of authors [33], dealing with broader topics.

Previously existing links between population geography and neighboring social, architectural, planning, and other disciplines were strengthened. As a result many of the concepts and methods of population geography were adopted by sociologists, economists, and planners. Population geography was, in its turn, considerably influenced by these disciplines, and research devoted to constructive goals increased. The general development of the population sciences gave rise to the necessity of defining the place occupied by population geography in this complex of disciplines [23]. Objections were expressed to the creation of a general science of population [69].

In population geography itself the thematic range of research is still fairly broad, although *urban geography* continues to predominate as a result of the practical requirements posed by rapid urbanization. This field, because of its relatively high level of development, is sometimes regarded not as a part of population geography but as a separate economic geographic discipline [110].

Publications dealing with synthesis, considering population and settlement a whole, are still few in number as a result of the trend towards specialization in research. Of considerable interest from this point of view are publications dealing with the regionalization of settlement, presenting both a new methodology and the results of the application of this methodology to the USSR [98, 99]. Another synthetic trend is research into territorial settlement patterns, such as local groupings of urban and rural settlements the various ways in which they interact [51, 114a].

Within the field of urban geography the differentiation of research trends is proceeding rapidly. Integrational tendencies are of course also appearing as a result of the attempt on the part of specialists in various areas to study cities as complex social phenomena. The possible emergence of a unified science of cities embracing all aspects of urban research has been discussed [233].

---

*A survey of the literature in population geography [173] from 1961 to 1965 covered approximately 2000 publications, while Khorev's bibliographic summary of work published during the period from 1955 to 1961 covered approximately 1100 publications.

On the basis of previously formulated theoretical conceptions and methods there appeared works of a general nature on the geography of cities that analyzed various aspects of the origination, growth, and structure of urban settlement in the USSR [233] and discussed the basic categories of this discipline, the problems involved in research and application of results, and its links with developments in city planning [110].

The intensified study of the specific types and forms of urban population (agglomerations, large cities, small cities, settlements of the urban type, etc.), as well as a strengthening of the links between urban geography and applied research on the allocation of productive resources and regional and urban planning, stimulated discussion of the ways in which cities of various types develop and of urban construction policy. This topic was covered at the Second Conference on Population Geography [9, 21, 111, 112]. The development of small- and medium-sized cities in the context of preplanning research on the allocation of productive resources, the formation of industrial blocks and agro–industrial complexes, i.e., in close association with regional research into the concrete conditions prevailing in the central regions, were studied by the Gosplan USSR Council for the Organization of Productive Resources [190]. Agreement was reached with regard to the fundamental issue of long-range development of Soviet cities: on the need to analyze and plan settlements in the form of a unified system, including interconnected elements of various types and performing various functions [232]. Differing views continued to be expressed, however, with regard to the role of various elements (agglomerations and large cities on the one hand and small cities on the other) and long-term city construction policy as regards these groups of elements.

In addition, it should be noted that, in the research described above, shortcomings previously characteristic of urban geography were eliminated: the tendency to descriptiveness, isolation from other trends in economic–geographic research and in neighboring urban-study disciplines, the consideration of single cities in isolation, and static analyses. Even now, however, insufficient attention is devoted to analyzing the developmental processes of systems of urban settlement, of the functioning of its various elements. Many recommendations, both those proclaiming the expediency of unlimited development of large cities and those which advocate limitations of their development, with a shifting of the center of gravity of urban construction policy to small settlements, are postulated on general considerations rather than research into the actual processes and principles of urban development.

Increased attention was devoted to the regionalizing role of cities and systems of urban settlement, i.e., their central functions, which previously had been inadequately studied as a result of concentration on the productive functions and specialization of cities [82, 157, 190 et al.]. In this way the links between population geography and research in regionalization were strengthened.

This link between urban-studies research and the study of the territorial organization of productive resources, including the concepts of the PTC and the energy-production cycle, appeared even more clearly in another, separate trend in urban geography — the study of urban settlements from the systems approach viewpoint with broad application of mathematical methods. This line of research, which began at Odessa University, is now continuing to develop at Kazan University through the efforts of a group of geographers and mathematicians [18, 19, 20, etc.].

During the period under consideration there appeared attempts at the geographical interpretation of urbanization processes, which were considered as part of a broad social and economic phenomenon rather than reduced simply to urban growth [110, 127, 171, 177].

The study of settlement patterns became the broadest field for the application of mathematical methods. Although the above-mentioned publications by representatives of the Kazan mathematical geography school present data on the application of mathematical tools to the most significant and, therefore, traditional problems of geographical urban studies, the new methods in many instances were associated with topics that are not widely studied in Soviet geography. One such topic was the quantitative study of the principles, of great importance in city planning, governing settlement within the limits of cities and agglomerations [51, 106, 194]. Considerable effort was devoted towards improving intracity settlement models developed abroad and towards developing original ones [38, 46, 47, 132, 133], as well as towards constructing theories to describe settlement configurations and determine the degree of their regularity [133]. At the same time, work continued on the elaboration of previously formulated concepts, in particular of the economic–geographic status of cities [123].

THE ECONOMIC–GEOGRAPHY STUDY OF NATURAL RESOURCES

Although Soviet economic geography has always emphasized the study of natural factors in industry, the period under consideration witnessed the development of a separate subsection of economic geography devoted to the appraisal of natural resources.* It is particularly in this sense that the thesis of an economic–geographic science of resources has come to the fore as a new trend.

The role of this line of research is all the more important in that it embodies the participation of economic geography in the defense and im-

---

*A rather comprehensive survey of the literature on this question is contained in two special editions of "Progress in Science – The Economic Appraisal of Natural Resources and Conditions of Production," Geografiya SSR (Geography of the USSR), No. 6, 1968, and "The Geography of Natural Resources," *ibid.*, No. 9, in press. For this reason the current discussion of this topic will be limited to a few short remarks.

provement of the environment, a task that is interdisciplinary in nature and which is attracting increasing attention both in the Soviet Union and abroad. At the same time, the development of research in this area strengthens the links between economic geography and the natural-scientific geographical disciplines and, consequently, facilitates the development of the system of the geographical sciences as a whole.

Since the theoretical roots of this new field extend to the philosophical *problem of the interaction of society and nature,* a problem which, however, has a clearly geographical aspect, many economic geographers have participated in its methodological elaboration. Important material on this subject was presented in the collection of articles *Nature and Society* (Moscow, Science Press, 1968), which consists primarily of somewhat reworked material from a conference of geographers and philosophers held in 1964. Questions concerning the influence of the geographical environments of society [186], the spheres of interaction of society and nature [55], the nature of natural resources [91, 139], etc., were comprehensively discussed. Many of the positions taken in the articles presented in this collection, however, both by economic geographers and specialists in other fields (natural scientists, philosophers, etc.), evoked sharp criticism [215, 161].

The problem of the interaction of society and nature, considered from the ethnogenetic point of view as a natural-historical phenomenon, received special attention [45 etc.]. This position was criticized by economic-geographers, among others [218].

The problem of the interaction between society and nature was posed in its general form from the viewpoint of population geography, and the creation of a new discipline studying the interaction of nature and human population, geodemology, was proposed [104, 105]. Problems involving the interaction of society and nature, as well as the concept of the geographical environment, attracted the attention of individual philosophers [119]. A number of interesting ideas relating to the place of these issues in science were expressed by Kolotiyevskiy [90].

Research in the area of the economic-geographic study of productive resources is developing along a number of lines.

An original set of calculations with serious methodological underpinnings established the quantitative parameters for the exploitation of natural resources, elucidating the structure, dynamics, and territorial (regional) characteristics of extraction and use of primary natural materials [92, 93]. This served as the basis for the concept of *resource cycles,* which characterizes the process of exploitation of natural substances as a cycle consisting not only of the extraction and processing of matter and energy, but also the formation of wastes and their ejection into the environment [95]. A methodological and methodical basis has thus been created for a systematic economic-geographic analysis of the entire process of natural-resource use that retains a continuous bond with the preservation of the environment against anthropogenic pollution.

In the area of the *economic appraisal of natural resources,* which in recent years has been of great interest to many economists, attempts were undertaken to create a generalized economic–geographic conception of these processes based on generalization of extensive branch-oriented research in appraising various types of natural resources, especially land and minerals. These attempts relied mainly on elaboration of the theory and methods of economic appraisal of natural resources, understood as the analysis of territorial variations in the efficiency in using these resources [138, 165]. A somewhat different course was followed by research on methods for appraising the natural conditions of life for human populations, including analysis not so much of economic as of social and medical–biological factors, involving primarily the use of the cartographical method for expressing appraisals [116, 118, 144, 240, etc.]. A theoretical and methodological generalization of data accumulated by geographical-appraisal research was carried out at a Working Conference on this subject held by the Institute of Geography of the Academy of Sciences of the USSR in 1970 [117]. An original technique for the quantitative appraisal of conditions relevant to agricultural reclamation, including a cartographical analysis of the areas of manifestation of many factors, was proposed for use in Eastern Siberia [101].

The links between the trend towards study of natural resources and main-stream economic geography are reflected in publications specially devoted to the analysis of the role and mechanism of action of natural factors in the development of the territorial structure of the economy, including the development of territorial economic complexes [140, 159, etc.].

REGIONAL ECONOMIC GEOGRAPHY OF THE USSR

Although regional economic geography (economic–geographic area studies), the scientific foundation of which was created by Baranskiy and his students, has for several decades been the center of gravity of economic–geographic research in the Soviet Union, the situation has now changed.

The major reason for this lies in the enormous importance attached to the multifaceted "deductions and knowledge" economic activity of the Soviet Union in relation to the natural, ethno–demographic, and historico–economic characteristics of the various regions that were resolved in the mid-1960s.*

This effort, which entailed the collection, processing, and analysis of an enormous amount of factual material, clarified, crystallized, and completed the basic directions and branches of Soviet economic geography. It was

---

*It is not the case that the task, formulated by Baranskiy, of creating a "Geography of the USSR has not been fulfilled. The periodical and nonperiodical regional publications of the 1950s and 1960s, at least on the scientific and methodological level, fulfilled this task. There are no longer any blank spots on the economic map of the USSR.

in regional research that the first generations of Soviet economic-geographers learned their profession, and the basic scientific and pedagogical working groups developed in academic institutions.

The more recent scientific and practical problems, many of which arose in the course of regional research, could not be solved within the scope of classical multiple regional characterizations, with their typically broad range of material "from geology to ideology."

Meanwhile the practical need for regional economic, cultural, and educational characterizations of various types, levels of detail, and expository forms remains unchanged. In addition, rapid changes in the objects being described — the regions and their individual parts — causes previous work to become obsolete and, therefore, gives rise to the need to periodically update them.

The completion of the scientific economic–geographic description of the regions of the USSR created the basis for the development of new periodical publications of various types. Among these publications are the economic–geographic sections of the academic series *Natural Conditions and Resources in the USSR*, the series of regional surveys published in the form of instructional manuals by the Prosveshchenyie Press, the Soviet Union series of popular scientific books on geography published on a large scale by the Mysl' Press, and the many books published by local publishing houses. Accordingly, it should be noted that, in regard both to methodology and characterization procedures, all these publications rely completely on the principles formulated by regional economic geography.

The situation with regard to monographs on regional topics based on preplanning research conducted in accordance with the General Scheme for Allocation of Productive Resources is somewhat different.* Written primarily by allocation economists working for the Gosplan USSR SOPS or under its direction, these publications reflect an approach to the study of regions that is different from traditional economic–geographic characterizations. The most general features of this group of publications maybe described as follows.

The problem-oriented analytic approach and the prediction of future developments predominate. Analysis of the branch structure of production is emphasized at the expense of detailed consideration of the territorial organization of productive resources. Much less attention than in economic–

---

*These publications, which do not as yet cover the entire territory of the USSR (books have come out on the North-West, Kazakhstan, Western and Eastern Siberia, and the Far East), are not formally issued as parts of a series, although in a number of respects they may be so regarded. The appearance of monographs on regional subjects preceded the publication of a collection of articles dealing with the central problems of allocation of the productive resources of all of the economic regions of the USSR [192]. The purpose of this collection is in many ways similar to that of the "Blue Series" collection '*Geographical Problems in the Development of the Major Economic Regions of the USSR* (Moscow, Mysl Press, 1964), which is now nearing completion. In the bibliography of SOPS (Council for the Organization of Productive Resources) collection, however, neither this collection nor any other economic–geographic monographs on regional topics are included.

geographic works is devoted to characterization of natural conditions and population, to the historico-genetic aspects of development of the current structure of the economy, the lowland regions, and the cities. These publications are replete with factual, including computational, data relevant to contemporary problems, but are not oriented toward the creation of a wholistic representation of the interaction of all of the region-forming factors or of the territorial structure of the entire economy of a region.

Although economic-geographic regional characterizations are based at the methodological level on the fully elaborated (primarily in the works of Baranskiy) principles and procedures of regional description, the publications in question are not based on any clearly formulated methodology. In many but not all respects they reflect the approaches used in the elaboration of the General Scheme [162].

Clearly, the absence of efforts to improve the methodology on which these projects are based is the result of a general decrease in interest in regional research. Interesting proposals on the standardization and formulation of regional characterizations [195] have evoked neither response nor practical application.

Multiple or complex-regional projects do not represent the mainstream of current economic-geographic research. Among the latter, application-oriented, relatively narrowly specialized studies predominate. A theoretical-methodological re-equipping is also detectable in this research — the use of the analytic approach in place of the observation and description, the use of quantitative methods and elements of modeling, etc. In much regional research there is an attempt to use concrete material to test current methodological conceptions and techniques.

The collections of articles published by Siberian Scientific institutes (*Reports of the Institute of Geography of Siberia and the Far East, Economic-Geographic Problems in the Development of Territorial-Productive Complexes in Siberia, Problems of Geography,* et al.) show these tendencies most clearly.

Regional problems attracting special attention include those having to do with the development of PTCs under various natural and economic conditions [22, 25, 67, 68, 134, 147, 219, etc.].

The continuing process, though changing in its technological and economic forms, of reclaiming new regions primarily in the North has been attracting considerable interest on the part of economic geographers. The period under consideration is characterized by a striving for the geographical generalization, including typological and classificatory generalization, from the enormous practical experience accumulated during the assimilation of territories in the European, Siberian, and Far-Eastern North [53, 54, 102, etc.].

The resource-appraisal trend has produced regional studies analyzing the role of natural factors in the development of various regional economic complexes [30].

THE ECONOMIC-GEOGRAPHIC STUDY OF FOREIGN COUNTRIES

The social and economic conditions prevailing in foreign countries, as well as their goals, conditions, and methods of research, all of which differ fundamentally from those prevailing in the Soviet Union, gave rise to the traditional development of economic geography in those countries as an independent branch of science. The development of a world socialist system and the liberation of tens of countries from colonial oppression in turn stimulated the development of the economic geography of socialist and developing countries. The general course of development of the economic geography of the capitalist countries was examined in a special survey published in the USSR [213].

The development of economic, scientific, and technological cooperation between the Soviet Union and foreign socialist and developing countries has significantly altered the direction of the economic-geographic study of these countries, causing it to assume a more applied character. By the same token increased opportunities developed for economic geographers to visit these countries, to collect required material at first hand, and to develop professional contacts and do cooperative research with foreign geographers.

The increased requirements of various organizations for scientific information regarding foreign regions and countries gave rise to the creation of a significant number of scientific centers of the regional geography type (associated with the Academy of Sciences of the USSR – Institutes for Study of the Economy of the World Socialist System, Africa, Latin America, the Far East, the United States, et al.). Although the range of problems faced by these countries is extremely broad, economic-geographic studies have proved a necessary element in the general pattern of research in regional studies. The organization of specialized institutes exerted both direct and indirect influence on the direction and content of economic-geographic studies carried out in other scientific centers – geographical and scientific research organizations as well as academic institutions.

The development of economic-geographic thought in the Soviet Union has also been an important factor in foreign research in this area. The general methodological positions and specific techniques developed by Soviet economic-geographers are being utilized abroad.* As a result, the purely regional geography approach – the compiling of complex characterizations of individual countries – that, until recently, has been overwhelmingly predominant in economic-geographic research, is being replaced by a specialized, problem-oriented regional approach. In this respect the tendencies observed in foreign economic geography correspond with those characteristic of

---

*Among economic geographers actively working on methodological problems a considerable number specialize in the study of foreign countries: Gokhman, Mayergoiz, Medvedkov, Semevskiy, etc.

Soviet economic geography.* Significant differences also exist, however,

Wide differences in the socio-economic and geographical conditions prevailing in different countries make it necessary to study concrete economic objects if generalizations concerning these countries are not to emerge as vulgar sociological schematizations. This situation in turn requires that investigators be sufficiently specialized, especially with regard to their linguistic and historiographic preparation. In most instances, therefore, generalizations deriving from research in this area are based on concrete data regarding individual countries and groupings of countries and, consequently, are taking on more or less clearly expressed regional characteristics.

The significant number of studies dealing with global economic problems or very general methodological issues in the economic-geographic study of foreign countries are clearly relevant in this context. Among these studies we may note a number dealing with extremely complex and important problems — the economic-geographic typology of the capitalist countries [28], the development of typological approaches in the economic-geographic study of socialist countries [250], and the comparison of various socio-economic types of the productive complex [3]. Of general methodological significance are studies elucidating the characteristics, directions, and problems in the geographical study of populations of capitalist and developing countries [174].

During the period under consideration there were only isolated attempts to analyze the geography of individual branches of production on the scale of the world economy [29, 120, 199, etc.]. As before, most studies deal with specific types and groups of countries.

The economic-geographic study of foreign socialist countries, in addition to continuing the traditional characterization of individual states, addressed itself to an increasing extent towards general problems relating to members of the Council for Mutual Economic Assistance (COMECON). This is understandable, in view of the growing and multifaceted political, economic, and cultural cooperation between the countries of the world socialist system and the development of their economic integration.

Characteristic manifestations of this tendency can be noted in major research efforts undertaken in the areas of fuel resources [128], energetics [124], and populations [170] of the European socialist countries. The principles of a typology of industrial regions [121] have been developed relative to this group of countries. Each of these studies not only makes wide use of the transition to a combined analysis of an entire group of countries and, consequently, to a higher level of generalization but also makes wide use of (and broadens and develops) the methodological achievements of Soviet economic geography.

---

*It should be noted, however, that these tendencies are as yet only weakly reflected in foreign geography textbooks, which are written basically on the level of traditional regional geography [17, 242, 243, etc.].

The transition to the economic-geographic interpretation of phenomena and processes associated with the current stage in the economic integration of socialist countries has appeared in response to the requirements of the modern age [126, etc.]. A significant portion of the scientific and, especially, applied research in the allocation sphere of the productive resources of foreign socialist countries is being done outside geographical institutions — primarily by economic organizations.

An extremely characteristic feature of the period under consideration lies in the concentration of the efforts of a significant number of economic geographers on *studying developing countries*. This is reflected in the appearance of regional geography and problem-oriented applied studies and in the organization of several major conferences dealing with this subject. The study of developing countries has proved to be the area in which collaboration between economic geography and neighboring disciplines — political economy, economics, ethnography, history, sociology, etc. — as well as between economic geographers and the organizations implementing scientific and technical cooperation between the USSR and the developing countries, has been most effective. As the result of a number of objective and subjective factors the most successful efforts of this sort have been the economic-geographic investigation of African and South Asian countries (especially India). Fundamental problems in the economic-geographic study of this group of countries have been formulated [131].

A number of themes have appeared in studying developing countries. The most characteristic theme, however, is the attempt to apply Soviet methodological conceptions to discover the essential features of problems and and principles governing the territorial structure of the economy of developing countries, and also to find ways of utilizing Soviet experience in the economic development in these countries. Such attempts are of interest in light of the Soviet policy of aid to countries that are following noncapitalist routes to the development of their national economies.

The specific features of the economies of developing countries are such that a large proportion of the economic-geographic problems with which they must deal are directly or indirectly associated with the use of natural resources. It is therefore hardly accidental that the trend toward study of third-world countries simultaneously implied increased attention to the study and appraisal of their resource potential.

The most actively developing area is the study of economic regionalization, which is closely associated with regional planning and improvement of the territorial structure of the economies of these countries. A special conference was devoted to this field, and the resulting data were later published [246].*

---

*The appearance of this collection of papers in English was especially timed to coincide with the 21st International Geographical Congress, which took place in 1968 in India.

Although monographs on regional geography published in the past were devoted to the isolation of economic regions and their more or less detailed description, those referred to above dealt with a wide range of methodological questions. In addition to presenting general theoretical formulations relating to the economic regionalization of developing countries [85], the monographs in question discussed a number of specific factors relating to natural resources [35, 200] and the historical legacy of the colonial period [187]. Many general problems passing beyond the regional framework were analyzed in the course of studies of the development of economic structure and regions in various parts of the world — India [211, 212, et al.], Africa [115, 229, et al.], and Latin America [129]. The problem of modeling territorial socio-economic systems was formulated [115a].

The extremely urgent problem of hunger, which constitutes a continuous threat to the greater part of the populations of developing countries, has stimulated efforts towards an economic-geographic interpretation of the most important social problem. Rejecting pessimistic neo-Malthusian conceptions, Soviet economic geographers assert the need for extensive, geographically differentiated investigation of this problem [83, 86, 189, etc.].

The results of this stage in the economic-geographic study of the developing countries were summarized at a conference held by the Geographical Society of the USSR and the Institute of Geography of the Academy of Sciences of the USSR [31].

Considerable attention has been attracted by population geography, including the study of the urbanization process. Experience in generalizing data derived from studies in the population geography of individual countries and regions was reflected in a speical collection of articles [33].*

Study of the developed capitalist countries, especially Western Europe and Japan, is proceeding less intensively, and is primarily problem-oriented. Of interest in this area is the critical study of foreign experience that could prove useful in the economic development both of populated and, especially, of newly settled regions of the USSR [5, 6]. Particular attention is being devoted to the general processes involved in the allocation of production at the present stage of development of these regions [42, etc.].

The most traditional line of research in this field — characterization of the regional geography type per se — is experiencing a gradual change in its normal descriptive procedures. This is reflected in changes in the relative lengths of various sections of publications and an abundance of problem-oriented discussions and typological schemes; in a word, in a strengthening of the analytical and constructive foundations of the field. In addition, the transition to a new type of regional geography can be observed, one which covers a major region and exposes and formulates major problems in such

---

*A collection of articles on problems of urbanization appeared somewhat earlier, i.e., before the period under consideration here [39].

areas as the use of natural resources, demographic explosions, urbanization, industrialization, and economic regionalization. This new approach is already being applied to Latin America [130].

At the same time the requirements of scientific and technical collaboration, tourism, and the teaching of geography are giving rise to a large number of regional geography publications of the reference and popular variety (folding maps published jointly by the Main Administration of Geography and Cartography of the Soviet of Ministers of the USSR and Mysl' Press, the series "At the Map of the World," et al.), as well as scientific economic-geographic characterizations of various countries in monograph form. Major popular science series ("Countries and Peoples of the World," etc.) are in preparation.

The following conclusions may be drawn from the brief survey presented above.

During the period in question Soviet economic geography achieved significant gains in the development of its theory and methods of investigation.

The subject matter covered in economic-geographic publications expanded, and individual disciplines and lines of research continued to develop. The differentiation of economic geography, which in the past has been somewhat slow, showed gradual development.

Integrational tendencies increased: links and continuous collaboration both with natural-scientific, especially geographical, sciences and with social and engineering disciplines were strengthened. The range of disciplines with which economic geography has established ties is widening and going beyond the scope of the economic sciences.

All of the new tendencies noted in the survey "coexist" with traditional conceptions and approaches, and this coexistence gives rise to the multiplicity of types of current economic-geographic studies, as well as sharp debate on the issues in question. There is therefore reason to consider this period as a *transitional* one, inasmuch as the future contours of Soviet economic geography are as yet far from completely defined.

It is clear even now, however, that objective conditions — the requirements of the current stage of the building of Communism in the Soviet Union and the influence of the scientific and technological revolution — are making rapid progress in the theory and practice of Soviet economic geography not only possible but necessary.

LITERATURE CITED

1. Aganbegyan, A.G. Mathematical-economic modelling in the solution of problems in the optimal allocation of productive resources, Izvestiya (Bulletin) of the Siberian Branch of the Academy of Sciences of the USSR, 1967, No. 6, Seriya Obshchaya Nauka, Issue 2.
2. Agafonov, N.T., Al'tman, L.P., Lavrov, S.B., Semevskiy, B.N., Chertov, L.G. New trends in the development of economic geography, Vestnik (Journal) of Leningrad State University, 1970, No. 18.
3. Agafonov, N.T. and Lavrov, S.B. Principle differences between capitalist and socialist regional productive-territorial complexes, Vestnik (Journal) of Leningrad State University, Leningrad, 1966, No. 6, Issue 1.
4. Agafonov, N.T. and Lavrov, S.B. Current problems in economic geography in light of Lenin's legacy, Vestnik (Journal) of Leningrad State University, 1970, No. 6.
5. Agranat, G.A. Foreign experience in assimilating northern territories, Izvestiya (Bulletin) of the Academy of Sciences of the USSR, Seriya Geografiya, 1967, No. 3.
6. Agranat, G.A. Northern Territories: Foreign experience in their assimilation (Zarubezhnyi Sever: opyt osvoyeniya), Moscow, Nauka Press, 1970.
7. Adamchuk, V.A. and Dvoskin, B.Ya. Problems in the development of industrial blocks in the USSR, as illustrated by Kazakhstan (Problemy razvitiya promyshlennykh uzlov SSSR na primere Kazakhstana), Moscow, Mysl' Press, 1968.
8. Alayev, E.B. Efficiency of the complex development of an economic region (Effektivnost' kompleksnogo razvitiya ekonomicheskogo rayona), Moscow, Nauka Press, 1965.
9. Alayev, E.B. and Khorev, B.S. Development of small and medium-sized cities in the USSR. In: Scientific problems in population geography (Nauchnyye problemy geografii naseleniya), Moscow, Moscow State University Press, 1967.
10. Alampiyev, P.M. Lenin's works and some problems in economic-geographic science, Izvestiya (Bulletin) of the Academy of Sciences of the USSR, Seriya Geografiya, 1970, No. 3.
11. Al'brut, M.I. Classics of Marxism-Leninism – an effective weapon in the struggle against the bourgeois theory of social physics and its geographical variety, Doklady (Reports) Chelyabinskogo Otdela Geograficheskogo Obshchestva SSSR, Leningrad, 1970.
12. Anuchin, V.A. Mathematization and geographical method, Vestnik (Journal) of Moscow State University, Seriya Geografiya, 1966, No. 6.
13. Anuchin, V.A. Problems of geography and goals in the propagation of geographical knowledge (O problemakh geografii i zadachakh propagandy geograficheskikh znaniy), Moscow, Znaniye Press, RSFSR, 1968.
14. Bandman, M.K. and Larina, N.I. Models of productive-territorial complexes in optimizing the allocation of production in an economic region, Voprosy geografii, Collection 80, Moscow, Mysl' Press, 1970.
15. Bandman, M.K. and Panchenko, A.I. Mathematical-economic models of the development of territorial-productive complexes, Izvestiya (Bulletin) of the Siberian Branch of the Academy of Sciences of the USSR, Seriya Obshchaya Nauka, 1967, No. 6, Issue 2.

16. Bandman, M.D. and Panchenko, A.I. Investigation of territorial-productive complexes with the aid of mathematical-economic modelling, Voprosy geografii, Collection 75, Moscow, Mysl' Press, 1968.
17. Barsov, N.N., Ginzburg, N.S., Dolinin, A.A. Lectures on the economic geography of foreign countries (Lektsii po ekonomicheskoi geografii zarubezhnykh stran), Part 1, Leningrad, Leningrad State University Press, 1967.
18. Blazhko, N.I. Quantitative methods of studying urban settlement systems. In: Geography of population and populated sections in the USSR (Geografiya naseleniya i naselennykh punktov SSSR), Leningrad, Nauka Press, 1967.
19. Blazhko, N.I., Voskoboynikova, S.M., Gurevich, B.L. Urban settlement systems. In: Scientific problems in population geography (Nauchnyye problemy geografii naseleniya), Moscow, Moscow State University Press, 1967.
20. Blazhko, N.I., Grigor'yev, S.V., Zabotin, Ya.I. Mathematical-geographic methods of studying urban settlements (Matematiko-geograficheskiye metody issledovaniya gorodskikh poselenii), Kazan', Kazan' State University Press, 1970.
21. Bogorad, D.I. Study and regulation of the growth of urban agglomerations. In: Scientific problems in population geography (Nauchnyye problemy geografii naseleniya), Moscow, Moscow State University Press, 1967.
22. Bud'kov, S.T. Sosvin region of the River Ob — a developing territorial-productive complex. In: Economic-geographical problems in the development of the territorial-productive complexes of Siberia (Ekonomiko-geograficheskiye problemy formirovaniya territorial'no-proizvodstvennykh kompleksov Sibiri), Novosibirsk-Irkutsk, 1970.
23. Valente, D.I. and Koval'skaya, N.Ya. The status of population geography in the system of the population sciences. In: Scientific problems in population geography (Nauchnyye problemy geografii naseleniya), Moscow, Moscow State University Press, 1967.
24. Valesyan, L.A. The productive-territorial complex of the Armenian SSR. Economic-geographical research into problems of formation and development (Proizvodstvenno-territorial'nyy kompleks Armyanskoi SSR. Ekonomiko-geograficheskoye issledovaniye problem formirovaniya i razvitie), Yerevan, Aiastan Press, 1970.
25. Varlamov, V.S. Development of the Ob-Irtysh regional productive complex, Voprosy geografii, Collection 80, Moscow, Mysl' Press, 1970.
26. Vashchenko, A.T. Problems of economic-geographical science in the works of V.I. Lenin. In: Economic geography. Interagency scientific collection (Ekonomicheskaya geografiya, Mizvidomehyy nauky zbornik, Issue 8, 1970.
27. Vetrov, A.S. Fundamental stages in the development of the methodology of economic regionalization. Economic regionalization as a method and as a science (Osnovnyye etapy razvitiya metodologii ekonomicheskogo rayonirovaniya. Ekonomicheskoye rayonirovaniye kak metod i kak nauka), Chelyabinsk, Yuzhnyy-Uralskiy Press, 1967.
28. Vol'sky, V.V. Types of countries in the capitalist world, Vestnik (Journal) of Moscow State University, Seriya Geografiya, 1968, No. 6.
29. Vol'f, M.B. Tendencies in the world production of agricultural products and in the

demand for food for the 1975-1985 period, Izvestiya (Bulletin) Vsesoyuznogo Geograficheskogo obshchestva, 1968, No. 5.
30. Geographical problems in the complex development of productive resources and the acquisition of the natural resources of the USSR (Geograficheskiye problemy kompleksnogo razvitiya proizvoditel'nykh sil i osvoyeniya yestestvennykh resursov SSSR), A collection of articles, Irkutsk, 1968.
31. Geography and the developing countries. Problems in the use of natural and labor resources (Geografiya i razvivayushchiyesya strany. Problemy ispol'zovaniya prirodnykh i trudovykh resursov), Abstracts of papers, Moscow, 1970.
32. Geography of population and populated sections in the USSR (Geografiya naseleniya i naselennykh punktov SSSR), A collection of articles, Leningrad, Nauka Press, 1967.
33. World population geography. In: Problems of geography (Voprosy geografii), Coll. 71, Moscow, Mysl' Press, 1966.
34. Gerasimov, I.P. Structural geography: Goals, methods, and results, Izvestiya (Bulletin), Vsesoyuznogo Geograficheskogo obshchestva, 1966, No. 5.
35. Gerasimov, I.P., Komar, I.V., Mashbits, Ya.G. Natural resources and the economic regionalization of the developing countries. In: Problems of Geography (Voprosy geografii), Coll. 76, Moscow, Mysl' Press, 1968.
36. Gerasimov, I.P., Mints, A.A., Preobrazhenskiy, V.S., Shelomov, N.P. Current geographical problems in the organization of recreation, Izvestiya (Bulletin) of the Academy of Sciences of the USSR, Seriya Geografiya, 1969, No. 4.
37. Gerasimov, I.P. and Preobrazhenskiy, V.S. Territorial aspects of the organization of the recreation and tourism industries. In: Development of the tourism industry (Razvitiye industrii turizma), Materials of a scientific conference, Novosibirsk, 1968.
37a. Golubitskaya, M.V. Interbranch balance of an industrial block. In: Problems in regional organization and efficiency of production (Problemy regional'noi organizatsii i effektivnosti proizvodstva), Moscow, 1970.
38. Gol'ts, G.A. Analytic expressions of patterns of population settlement. In: Mathematical methods in urban construction (Matematicheskiye metody v gradostroitel'stve), Kiev, Budyvel'nyk Press, 1966.
39. Cities of the world. In: Problems of geography (Voprosy geografii), Co.. 66, Mysl' Press, 1965.
40. Gokhman, V.M., Gurevich, B.L., Saushkin, Yu.G. Problems of metageography. In: Problems of geography (Voprosy geografii), Coll. 77, Moscow, Mysl' Press, 1968.
41. Gokhman, V.M. and Karpov, L. Cities and the allocation of production, Mirovaya ekonomika i mezhdunarodnaya otnosheniya, 1970, No. 3.
42. Granberg, A.G. Model of the territorial-productive complex in the system of models of the optimal development and allocation of individual branches of production. In: Problems in the optimal planning of the allocation of production (Problemy optimal'nogo planirovannogo razmeshcheniya proizvodstva), Novosibirsk, 1965.
43. Granberg, A.G. Models of production allocation in an optimal planning system, Investiya (Bulletin) of the Sibirian Branch of the Academy of Science of the USSR, 1967, No. 6, Seriya Obshchaya Nauka, Issue 2.
44. Granik, G.I. and Gromov, V.I. Branch and territorial distribution of labor (Otraslev-

oye i territorial'noye razdeleniye truda), Moscow, Mysl' Press, 1970.
45. Gumilev, N.L. Relation between nature and society according to historical geography and ethnography, Vestnik (Journal) of Leningrad State University, Seriya Geografiya, 1970, No. 24.
46. Gurevich, B.L. Urban population density and the density of a stochastic quantity, Vestnik (Journal) of Moscow State University, Seriya Geografiya, 1967, No. 1.
47. Gurevich, B.L. and Saushkin, Yu.G. Mathematical methods in geography, Vestnik (Journal) of Moscow State University, Seriya Geografiya, 1966, No. 1.
48. Guseva, V.D. Basic information requirements in modelling the territorial-productive structure of economic systems. In: Problems of geography (Voprosy geografii), Coll. 75, Moscow, Mysl' Press, 1970.
49. Guseva, V.D. and Kazanskiy, N.N. Territorial economic planning and complexes of productive resources. In: Productive-territorial complexes. Geography of the USSR (Proizvodstvenno-territorial'nyye kompleksy, Geografii SSSR), Results in Science (Itogi Nauki), Moscow, All-Union Institute of Scientific and Technical Information of the Academy of Sciences of the USSR, 1970, Issue 8.
50. Dzhaoshvili, V.S. Population of Georgia. An economic-geographical investigation (Naseleniye Gruzii. Ekonomiko-geograficheskiye issledovaniye), Tbilisi, Metsniyereba Press, 1968.
51. Davidovich, V.G. Territorial settlement systems in the USSR. In: Scientific problems in population geography (Nauchnyye problemy geografii naseleniya), Moscow State University Press, 1967.
52. Davidovich, V.G. Quantitative patterns of settlement relative to work location. In: Settlement in cities (Rasseleniye v gorodakh), Moscow, Mysl' Press, 1968.
53. D'yakonov, F.V. Theoretical and methodological aspects of the formation of the productive-territorial structure of the northern regions of Western Siberia. In: Natural conditions and characteristics of the economic reclamation of the northern regions of Western Siberia (Prirodnyye usloviya i osobennosti khozyaistvennogo osvoyeniya severnykh rayonov Zapadnoi Sibiri), Moscow, Nauka Press, 1969.
54. D'yakonov, F.V. Typology of the newly reclaimed northern regions of the USSR. In: Problems of Geography (Voprosy geografii), Coll. 80, Moscow, Mysl' Press, 1970.
55. Zhirmunskiy, M.M. Areas of interaction of nature and society and the relation in this interaction of natural and social elements in the geographical environment. In: Nature and Society (Priroda i obshchestvo), Moscow, Nauka Press, 1968.
56. Zaitsev, I.F. Territorial model of productive resources, Voprosy geografii, Coll. 77, Moscow, Mysl' Press, 1968.
57. Zaitsev, I.F. Classification of models in economic geography, Vestnik (Journal) of Moscow State University, Seriya Geografiya, 1970, No. 2.
58. Regularities and factors in the development of the economic regions of the USSR (Zakonomernosti i faktory razitiya ekonomicheskikh rayonov SSSR), Moscow, Nauka Press, 1965.
59. Principles of productive-territorial complexes. Abstracts of papers presented at an interrepublic scientific conference (Zakonomernosti formirovaniya proizvodstvenno-territorial'nykh kompleksov), Kiev, Naukova Dumka Press, 1968.
60. Zvonkova, T.V. and Saushkin, Yu.G. Problems of long-term geographical prediction,

Vestnik (Journal) of Moscow State University, Seriya Geografiya, 1968, No. 4.
61. Zvonkova, T.V. and Saushkin, Yu.G. Significance of the works of V.I. Lenin for scientific prediction, Vestnik (Journal) of Moscow State University, Seriya Geografiya, 1970, No. 2.
62. Zorin, I.V. Economic-geographical factors in the formation of recreational regions. In: Geographical problems in the organization of recreation and tourism. Abstracts of papers presented at a working conference (Geograficheskiye problemy organizatsii otdykha i turizma), Moscow, 1969.
63. Ivanov, K.I. Productive-territorial complexes in agriculture, Vestnik (Journal) of Moscow State University, Seriya Geografiya, 1967, No. 1.
64. Ivanov, K.I. Social production as a territorial-system formation as illustrated by agriculture, Izvestiya (Bulletin) of the Academy of Sciences of the USSR, Seriya Ekonomika, 1970, No. 2.
65. Ivanov, N.V. Methods of appraising the level of social services, Uchennyye zapiski (Scientific notes) of Perm' State University, 1970, No. 211.
66. Economic-geographical status and the allocation of social production. In: Problems of the development and allocation of social production (Voprosy razvitiya i razmeshcheniya obshchestvennogo proizvodstva), Rostov-on-Don, 1967.
67. Ionova, V.D. and Malinovskaya, M.A. Specific conditions of formation of the Sayansk territorial-productive complex. In: Economic-geographical problems in the formation of territorial-productive complexes of Siberia (Ekonomiko-geograficheskiye problemy formirovaniya territorial'no-proizvodstvennykh kompleksov Sibiri), Novosibirsk, 1969.
68. Irkutsk-Cheremkhovsk industrial region. Problems in the geographical study of the territory (Irkutsko-Cheremkhovskiy promyshlennyy rayon. Voprosy geograficheskogo izucheniya territorii), Irkutsk, 1969.
69. Kabulov, B.A. Some controversial issues in population geography, Uchennyye zapiski (Scientific notes) of the South-Osetinsk State Pedagogical Institute, Seriya fiziko-matematicheskaya i biologiya nauka, 1968, No. 13.
70. Kazanskiy, N.N. Economic-geographical aspects of a unified transport network in the USSR. In: Problems of geography (Voprosy geografii), Coll. 75, Moscow, Mysl' Press, 1968.
71. Kazanskiy, N.N. and Stepanov, P.N. Lenin and the territorial organization of the productive resources of the USSR. In: Problems in geography (Voprosy geografii), Coll. 80, Moscow, Mysl' Press, 1970.
72. Kalashnikova, T.M. Region-forming factors, Vestnik (Journal) of Moscow State University, Seriya Geografiya, 1967, No. 3.
73. Kalashnikova, T.M. Economic regionalization (Ekonomicheskoye rayonirovaniye), lectures, Moscow, Moscow State University Press, 1969.
74. Kalashnikova, T.M. Current problems in the theory of economic regionalization in light of Lenin's ideas, Vestnik (Journal) of Moscow State University, Seriya Geografiya, 1970, No. 2.
75. Kalashnikova, T.M. Productive-territorial complex as a complex territorial system: An instructional handbook (Proizvodstvenno-territorial'nyy kompleks kak slozh-

naya territorial'naya sistema. Uchebno-metodologicheskoye posobiye), Moscow, Moscow State University Press, 1970.
75a. Kalashnikova, T.M. Lenin's ideas on territorial planning and their significance for the solution of current problems in economic geography, Vestnik (Journal) of Moscow State University, Seriya Geografiya, 1970, No. 6.
76. Kalesnik, S.V. Significance of Lenin's ideas for Soviet geography. In: Materials of the Fifth Conference of the Geographical Society of the USSR (Materialy V s"ezda Geograficheskogo Obshchestva SSSR), Leningrad, 1970.
77. Kalesnik, S.V. Some misunderstandings in the theory of Soviet geography, Izvestiya (Bulletin) Vsesoyuznogo Geograficheskogo Obshchestva, 1971, No. 1.
78. Cartographical provision for economic development plans. Materials of a symposium held at the Third Scientific and Technological Conference on Cartography (Kartograficheskoye obespecheniye planov razvitiya narodnogo khozyaystva), Irkutsk, 1968.
79. Fundamental issues in the methodology of the fractional regionalization of the territory of a major economic region. In: Materials of the Moscow Branch of the Geographical Society of the USSR. Economic Geography (Materialy Moskovskogo filiala Geograficheskogo obshchestva SSSR. Ekonomicheskogo geografiya, Moscow, 1968, Issue 1.
80. Kibal'chich, O.A. Current state and practical aspects of the study of productive-territorial complexes. In: Productive-territorial complexes. Geography of the USSR (Proizvodstvenno-territorial'nykh kompleksy. Geografiya SSSR), Itogi nauki (Results in science), Moscow, All-Union Institute of Scientific and Technical Information of the Academy of Sciences of the USSR, 1970, Issue 8.
81. Kistanov, V.V. Complex development and specialization of the economic regions of the USSR (Kompleksnoye razvitiye i spetsializatsiya ekonomicheskikh rayonov SSSR), Moscow, Nauka Press, 1968.
82. Knobel'sdorf, E.F. Region-forming role of cities and major rural settlements. In: Geography of population and populated sections in the USSR (Geografiya naseleniya i naselennykh punktov SSSR), Leningrad, Nauka Press, 1967.
83. Knyazhinskaya, L.A. Growth of population and the use of productive resources in the developing countries as illustrated by countries in South and Southeast Asia, Izvestiya (Bulletin) Vsesoyuznogo Geograficheskogo Obshchestva), 1968, No. 5.
84. Kovalev, S.A. and Pokshishevskiy, V.V. Population and service geography. In: Scientific problems of population geography (Nauchnyye problemy geografii naseleniya), Moscow, Moscow State University Press, 1967.
85. Kovalevskiy, V.P. Economic regionalization of the developing countries. In: Problems of geography (Voprosy geografii), Coll. 76, Moscow, Mysl' Press, 1968.
86. Kovalevskiy, V.P. Humanity and productive resources (Chelovechestvo i prodovol'stvennyye resursy), Priroda Press, 1969, No. 8.
87. Kozlov, L.A. Optimal planning of the development and allocation of branches of industry. Methodological problems (Optimal'noye planirovaniye razvitiya i nazmeshcheniya otrasley promyshlennosti. Voprosy metadologii i metodiki), Novosibirsk, Nauka Press, 1970.

88. Kolosovskiy, N.N. Theory of economic regionalization (Teoriya ekonomicheskogo rayonirovaniya), Moscow, Mysl' Press, 1970.
89. Kolosovskiy, N.N. Theoretical problems in the economic regionalization of the USSR. In: Problems of geography (Voprosy geografii), Coll. 80, Moscow, Mysl' Press, 1970.
90. Kolotiyevskiy, A.M. Problems in the theory and methodology of economic regionalization as related to the general theory of economic geography (Voprosy teorii i metodiki ekonomicheskogo rayonirovaniya v svyazi s obshchei teoriyei ekonomicheskoi geografii), Riga, Zinatne Press, 1967.
91. Komar, I.V. Applied significance of geography and the practical research of Moscow geographers, Doklady (Reports) Instituta Geografii Sibirii i Dal'nogo Vostoka, 1965, Issue 9.
92. Komar, I.V. Dynamics and structure of the use of natural resources in the USSR, Izvestiya (Bulletin) of the Academy of Sciences of the USSR, Seriya Geografiya, 1968, No. 3.
93. Komar, I.V. Territorial progress in the use of the natural resources of the USSR during the Soviet period, Izvestiya (Bulletin) of the Academy of Sciences of the USSR, Seriya Geografiya, 1967, No. 5.
94. Komar, I.V. Geographical sciences in the transformation of natural into socially productive resources. In: Nature and society (Priroda i obshchestvo), Moscow, Nauka Press, 1968.
95. Komar, I.V. Exchange of substances between nature and society and its optimization, Izvestiya (Bulletin) of the Academy of Sciences of the USSR, Seriya Geografiya, 1969, No. 5.
96. Komar, I.V. and Mintz, A.A. Leninist principles of the allocation of productive resources and the development of the economic regions of the USSR, Izvestiya (Bulletin) of the Academy of Sciences of the USSR, Seriya Geografiya, 1970, No. 2.
97. Konstantinov, O.A. Economic geography in the USSR at the 50th anniversary of Soviet power, Izvestiya (Bulletin) Vsesoyuznoga Geograficheskogo Obshchestva, 1967, No. 5.
98. Konstantinov, O.A. Regionalization of settlement in the USSR. In: Scientific problems of population geography (Nauchnyye problemy geografii naseleniya), Moscow, Moscow State University Press, 1967.
99. Konstantinov, O.A. Determining settlement regions in the USSR. In: Geography of population and population centers in the USSR (Geografiya naseleniya i naselennykh punktov SSSR), Leningrad, Nauka Press, 1967.
100. Korzhov, N.I. System of taxonomic units for an integral economic regionalization of the USSR. In: Nauchnyye Zapiski (Scientific notes) Voronezhskogo otdel' Geograficheskogo obshchestva SSSR, Voronezh, 1967.
101. Kosmachev, K.P. Quantitative appraisal of specifications of local conditions, Doklady (Reports) Instituta Geografii Sibiri i Dal'nogo Vostoka, 1967, Issue 16.
102. Kosmachev, K.P. Economic reclamation of territory as the object of economic-geographical research, Doklady (Reports) Instituta Geografii Sibiri i Dal'nogo Vostoka, Irkutsk, 1969, Issue 24.

103. Kossov, V.V. Interbranch balance (Mezhotraslevoi balans), Moscow, Ekonomika Press, 1966.
104. Kotel'nikov, V.L. Population and nature. Formulation of the problems of geographical analysis, Uchenyye Zapiski (Scientific notes) of Moscow State Lenin Pedagological Institute, 1970, No. 297.
105. Kotel'nikov, V.L. and Saushkin, Yu.G. Population and nature. In: Scientific problems of population geography (Nauchnyye problemy geografii naseleniya), Moscow, Moscow State University Press, 1967.
106. Kudryavtsev, O.K. Mathematical methods in the study of settlement. In: Settlement in cities (Rasseleniye v gorodakh), Moscow, Mysl' Press, 1968.
107. Kurakin, A.F. Territorial differences in conditions as the subject matter of social geography. Theoretical-methodological problems, Uchenyye Zapiski (Scientific notes) of Perm' State University, 1970, No. 211.
108. Lavrishchev, A.N. Administrative-territorial division of the USSR and economic regionalization. In: Economic geography (Ekonomika geografiya), Moscow, 1969.
109. Lavrov, S.B. New methods or new principles, Vestnik (Journal) of Leningrad State University, 1971, No. 1, Issue 6.
110. Lappo, G.M. Geography of cities and fundamentals of city planning (Geografiya gorodov s osnovami gradostroitel'stva), Moscow, Moscow State University Press, 1969.
111. Lappo, G.M. and Listengurt, F.M. Development of cities of various types in the USSR. In: Scientific problems of the geography of settlement (Nauchnyye problemy geografii naseleniya), Moscow, Moscow State University Press, 1967.
112. Lappo, G.M. and Troitskaya, Ye.Kh. Development of major cities in the USSR. In: Scientific problems of population geography (Nauchnyye problemy geografii naseleniya), Moscow, Moscow State University Press, 1967.
113. Lasis, Yu.V. Quantitative characterization of the typological features of the regional productive complexes of the USSR. In: Problems of geography (Voprosy geografii), Coll. 70, Moscow, Mysl' Press, 1970.
114. Leyzerovich, Ye.Ye. Significance of detailed economic regionalization for regional planning, Izvestiya (Bulletin) of the Academy of Sciences of the USSR, Seriya Geografiya, 1968, No. 1.
115. Lipetz, Yu.G. Macroeconomic regionalization of Africa. In: Problems of geography (Voprosy geografii), Coll. 76, Moscow, Mysl' Press, 1968.
115a. Lipetz, Yu.G. General problems of modelling the territorial social-economic systems of the developing countries. In: Analysis and prediction of the economies of developing countries (Analiz i prognozirovanniye ekonomiki razvivayushchikh stran), Moscow, 1970.
116. Lopatina, Ye.B. Selection of criteria and indicators for appraising natural living conditions. In: Problems of geography (Voprosy geografii), Coll. 78, Moscow, Mysl' Press, 1968.
117. Lopatina, Ye.B., Mintz, A.A., Mukhina, L.I., Nazarevskiy, O.R., Preobrazhenskiy, V.S. Theory and methodology of appraisal of natural conditions and resources, Izvestiya (Bulletin) of the Academy of Sciences of the USSR, Seriya Geografiya, 1970, No. 4.

118. Lopatina, Ye.B. and Nazarevskiy, O.R. Regional complex economic appraisal of natural conditions and resources, Izvestiya (Bulletin) of the Academy of Sciences of the USSR, Seriya Geografiya, 1966, No. 1.
119. Lyamin, V.S. Two levels of interaction between nature and society, Vestnik (Journal) of Moscow State University, Seriya Filosofiya, 1967, No. 3.
120. Mayergoyz, I.M. Geographical study of the world chemical industry. In: Problems of geography (Voprosy geografii), Coll. 72, Moscow, Mysl' Press, 1967.
121. Mayergoyz, I.M. Typology of the socialist countries of Europe, Vestnik (Journal) of Moscow State University, Seriya Geografiya, 1967, No. 6.
122. Mayergoyz, I.M. Foundations of the study of economic-geographic status. In: Soviet geographers attending the 21st International Geographical Congress (Sovetskiye geografy XXI Mezhdunarognogo geograficheskogo kongressu), Moscow, Nauka Press, 1968.
123. Mayergoyz, I.M. Some problems in the study of the economic-geographic status of the cities of the USSR. In: Materials of the Second Interagency Conference on Population Geography (Materialy Vtorogo mezhduvedennogo soveshchaniya po geografii naseleniya), Moscow, 1968, Issue 1.
124. Mayergoyz, I.M. Geography of the energy resources of the socialist countries of Europe (Geografiya energetiki sotsialisticheskikh stran zarubezhnoy Evropy), texts of lectures, Moscow, 1969.
125. Mayergoyz, I.M. Current stage in the development of Soviet economic cartography. In: New developments in the subject matter, content, and methods of compiling economic maps (Novoye v tematike, soderzhanii i metodakh sostavleniya ekonomicheskikh kart), Moscow, 1970.
126. Mayergoyz, I.M. Some territorial-geographical aspects of the economic integration of the socialist countries, Vestnik (Journal) of Moscow State University, Seriya Geografiya, 1970, No. 4.
127. Mayergoyz, I.M., Gokhman, V.M., Lappo, G.M., Pivovarov, Yu.L. Geographical aspects of urbanization. In: Problems of urbanization in the USSR (Problemy urbanizatsii v SSSR), Moscow, Moscow State University Press, 1971.
128. Maksakovskiy, V.P. Fuel resources of the socialist countries of Europe (Toplivnyye resursy sotsialisticheskikh stran Yevropy), Moscow, Nedra Press, 1967.
129. Mashbits, Ya.G. Formation of economic regions and economic regionalization of the countries of Latin America. In: Problems of geography (Voprosy geografii), Coll. 76, Moscow, Mysl' Press, 1968.
130. Mashbits, Ya.G. Latin America. Problems of economic geography (Latinskaya Amerika. Problemy ekonomicheskoy geografii), Moscow, Mysl' Press, 1969.
131. Mashbits, Ya.G. Problems in the economic-geographical study of the developing countries. In: Geography and the developing countries (Geografiya i razvivayushchiyesya strany), Moscow, 1970.
132. Medvedkov, Yu.V. Economic-geographic study of the regions of the capitalist world, Issue 2, Mathematics in Economic Geography. Results in Science (Itogi Nauki), All-Union Institute of Scientific and Technical Information of the Academy of Sciences of the USSR, 1965.
133. Medvedkov, Yu.V. Economic-geographic study of the regions of the capitalist

world, Issue 3, Analysis of settlement configurations. Results in science (Itogi Nauki), All-Union Institute of Scientific and Technical Information, Moscow, 1966,
134. Medvedkova, E.A. Development of the Irkutsk-Cheremkhovsk territorial-productive complex. In: Economic-geographic problems in the development of the territorial-productive complexes of Siberia (Ekonomiko-geograficheskiye problemy formirovaniya territorial'no-proizvodstvennykh kompleksov Sibiri), Novosibirsk, 1969, Issue 1.
135. Small-scale maps for the appraisal of natural conditions. Content, principles and methods of compilation (Melkomasshtabnyye karty otsenki prirodnykh usloviy. Soderzhaniye, printsipy i metody razrabotki), A collection of articles, Moscow, Moscow State University Press, 1970.
136. Merkusheva, L.A. Determining the level of population services. In: Economic-geographical problems in the development of the territorial-productive complexes of Siberia (Ekonomiko-geograficheskiye problemy formirovaniya territorial'noproizvodstvennykh kompleksov Sibiri), Novosibirsk–Irkutsk, 1970, Issue 2.
137. Methodological aspects of optimal branch planning in industry (Metodicheskiye polozheniya po optimal'nomu otraslevomu planirovaniyu v promyshlennosti), Novosibirsk, Nauka Press, 1967.
138. Mintz, A.A. Economic appraisal of natural resources and conditions of production. Geography of the USSR, Issue 6, Results in science (Itogi Nauki), All-Union Institute of Scientific and Technical Information, Moscow, 1968.
139. Mintz, A.A. Natural resources. The concept and some problems of classification. In: Nature and society (Priroda i obshchestvo), Moscow, Nauka Press, 1968.
140. Mintz, A.A. Natural resources as a factor in the formation of territorial-productive complexes. In: Problems of geography (Voprosy geografii), Coll. 80, Moscow, Mysl' Press, 1970.
141. Mintz, A.A. and Preobrazhenskiy, V.S. Place function and its variations, Izvestiya (Bulletin) of the Academy of Sciences of the USSR, Seriya Geografiya, 1970, No. 6.
142. Modelling the development and allocation of the production of branches of industry (Modelirovaniye razvitiya i razmeshcheniya proizvodstva otraslei promyshlennosti), Parts 1-2, Novosibirsk, 1970.
143. Mukomel', I.F. Economic regionalization of agriculture and its territorial organization. In: Problems of geography (Voprosy geografii), Coll. 75, Moscow, Mysl' Press, 1968.
144. Nazarevskiy, O.R. Selection of elements of the natural geographical environment and aspects of human life. In: Problems of geography (Voprosy geografii), Coll. 78, Moscow, Mysl' Press, 1968.
145. Nazarevskiy, O.R. and Zorin, I.V. Current problems in the geography of recreation and tourism. In: Development of the tourism industry (Razvitiye industrii turizma), Materials of a scientific conference, Novosibirsk, 1968.
146. Scientific problems in population geography (Nauchnyye problemy geografii naseleniya), Materials of the Second Interagency Conference on Population Geography, Moscow, Moscow State University Press, 1967.
147. Nedeshev, A.A. Basic stages in the formation of the Eastern Transbaikal territorial-

productive complex. In: Problems of geography (Voprosy geografii), Coll. 80, Mysl' Press, 1970.
148. Nekrasov, N.N. Scientific problems in the elaboration of a general scheme for the allocation of the productive resources of the USSR, Voprosy Ekonomiki, 1966, No. 9.
149. Nekrasov, N.N. Leninist conception of the efficient allocation of socialist productive resources, Vestnik (Journal) of the Academy of Sciences of the USSR, 1969, No. 11.
150. Nekrasov, N.N. Scientific bases of the general scheme for the allocation of the productive resources of the USSR through 1980. Summary of a report to the Fifth Conference of the Geographical Society of the USSR, Moscow, Council for the Study of Productive Resources, 1970.
151. Nekrasov, N.N. Problems in the allocation of the productive resources of the USSR, Vestnik (Journal) of the Academy of Sciences of the USSR, 1971, No. 3.
152. Nikotayev, S.A. Interregional and intraregional analysis of the allocation of productive resources (Mezhrayonnyy i vnutrirayonnyy analiz razmeshcheniya proizvoditel'- nykh sil), Moscow, Nauka Press, 1971.
153. Nikol'skiy, I.V. Role of branches of the economy in the formation of regional productive complexes, Vestnik (Journal) of Moscow State University, Seriya Geografiya, 1970, No. 2.
154. New developments in the subject matter, content and methods of economic map compilation (Novoye v tematike, soderzhanii i metodakh sostavleniya ekonomicheskikh kart), A collection of articles, Moscow, 1970.
155. Nutenko, L.Ya. Mathematical method for dividing territory, Vestnik (Journal) of Moscow State University, Seriya Geografiya, 1968, No. 5.
156. Nutenko, L.Ya. Theoretical-informational approach to the regionalization of territory, Vestnik (Journal) of Moscow State University, Seriya Geografiya, 1969, No. 4.
157. Nymmik, S.Ya. Regional systems of settlements as a region-forming framework, Vestnik (Journal) of Moscow State University, Seriya Geografiya, 1969, No. 3.
158. Nymmik, S.Ya. Nature of fractional economic regionalization, Uchenyye Zapiski (Scientific notes) of Tartu State University, Issue 242, Trudy (Transactions) po geografii, UP, Tartu, 1969.
159. Nymmik, S.Ya. Natural influences on the formation of social-economic territorial complexes, Vestnik (Journal) of Moscow State University, Seriya Geografiya, 1969, No. 1.
160. Nymmik, S.Ya. Region-forming centers, Vestnik (Journal) of Moscow State University, Seriya Geografiya, 1970, No. 1.
161. Discussion of Nature and Society (Vsesoyuznogo Geograficheskogo Obshchestva), A collection of articles, Izvestiya (Bulletin) of the Academy of Sciences of the USSR, Seriya Geografiya, 1969, No. 6.
162. General methods for elaborating the general scheme for the allocation of the productive resources of the USSR for the period 1971-1980 (Obshchaya metodika razrabotki General'noi skhemy razmeshcheniya proizvoditel'nykh sil SSSR na period 1971-1980), Moscow, Ekonomika Press, 1966.
163. General economic maps: content and methods of compilation (Obshcheékonomi-

cheskiye karty soderzhaniye i metody sozdaniya), A collection of articles, Moscow, 1969.
164. Optimal territorial-productive planning (Optimal'noye territorial'no-proizvodstennoye planirovaniye), Novosibirsk, Nauka Press, 1969.
165. Appraisal of natural resources. In: Problems of geography (Voprosy geografii), Coll. 78, Moscow, Mysl' Press, 1968.
166. Pavlenko, V.F. Territorial distribution of labor and regional economics, Izvestiya (Bulletin) of the Academy of Sciences of the USSR, Seriya Ekonomika, 1971, No. 2.
167. Pavlov, Yu.M. Regional politics of capitalist states (Regional'naya politika kapitalisticheskikh gosudarstv), Moscow, Nauka Press, 1970.
168. Parkhomenko, I.I. Significance of Lenin's "Sketch of a plan for scientific and technical work" for the study of the natural productive resources of the USSR. In: Geographical collection (Geograficheskiy sbornik), All-Union Institute of Scientific and Technical Information, 1970, No. 4.
169. Petryakov, T.I. Compilation of maps for the economic appraisal of natural resources, Izvestiya (Bulletin) of the Academy of Sciences of the USSR, Seriya Geografiya, 1970, No. 1.
170. Pivovarov, Yu.L. Population of the socialist countries of Europe. Structural-geographical developments (Naseleniye sotsialisticheskikh stran zarubezhnoi Yevropy. Strukturno-geograficheskiye sdvigi), Moscow, Nauka Press, 1970.
171. Pivovarov, Yu.L. Urbanization and the territorial structure of the economy. In: Urbanization and the working class in the scientific and technological revolution (Urbanizatsiya i rabochii klass v usloviyakh nauchno-tekhnicheskoi revolyutsii), Moscow, 1970.
172. Pokshishevskiy, V.V. Economic regionalization of the USSR. A survey of Soviet research on economic regionalization 1962-1964. In: Geography of the USSR (Geografiya SSSR), Results in Science (Itogi nauki), Issue 2. Economic regionalization of the USSR (Ekonomicheskoye rayonirovaniye SSSR), Moscow, All-Union Institute of Scientific and Technical Information, 1965.
173. Pokshishevskiy, V.V. Population geography in the USSR. In: Geography of the USSR (Geografiya SSSR). Results in science (Itogi Nauki), Issue 3, Moscow, All-Union Institute of Scientific and Technical Information, 1966.
174. Pokshishevskiy, V.V. Basic problems in the study of the geography of the populations of the capitalist and developing countries. In: Problems of geography (Voprosy geografii), Coll. 71, Moscow, Mysl' Press, 1966.
175. Pokshishevskiy, V.V. Geography of populations occupied in nonmaterial production and services. In: Geography of population and population centers of the USSR (Geografiya naseleniya i naselennykh punktov SSSR), Leningrad, Nauka Press, 1967.
176. Pokshishevskiy, V.V. Population as a productive resource and its geographical variations, Izvestiya (Bulletin) of the Academy of Sciences of the USSR, Seriya Geografiya, 1970, No. 2.
177. Pokshishevskiy, V.V. and Gokhman, V.M. Problems of hyperurbanization in the developed capitalist countries and its geographical aspects. In: Scientific problems of population geography (Nauchnyye problemy geografii naseleniya), Moscow, Moscow State University Press, 1967.

178. Pokshishevskiy, V.V., Mintz, S.A., Konstantinov, O.A. New trends in the development of Soviet economic geography. In: Materials of the Fifth Congress of the Geographical Society of the USSR (Materialy V s"ezda Geograficheskogo obshchestva SSSR), Leningrad, 1970.
179. Preobrazhenskiy, V.S. and Vedenin, Yu.A. Geography and recreation (Geografiya i otdykh), Moscow, Znaniye Press, 1971.
180. Privalovskaya, G.A. Characteristics of the geography of branches of industry engaged in the processing of primary natural materials, Izvestiya (Bulletin) of the Academy of Sciences of the USSR, Seriya Geografiya, 1970, No. 6.
181. Mathematical methods in the allocation of production (Primeniye matematicheskikh metodov razmeshcheniya proizvodstva), Moscow, Nauka Press, 1968.
182. Economic efficiency of the allocation of socialist production in the USSR (Problemy ekonomicheskoy effektivnosti razmeshcheniya sotsialisticheskogo proizvodstva v SSSR), Moscow, Nauka Press, 1968.
183. Probst, A.Ye. Efficiency of the territorial organization of production. Methodological outline (Effektivnost' territorial'noi organizatsii proizvodstva. Metodologicheskiye ocherki), Moscow, Mysl' Press, 1965.
184. Probst, A.Ye. Allocation of socialist industry (Voprosy razmeshcheniya sotsialisticheskoy promyshlennosti), Moscow, Nauka Press, 1971.
185. Productive-territorial complexes. Geography of the USSR (Proizvodstvenno-territorial'nyye kompleksy. Geografiya SSSR). Results in Science (Itogi nauki), Moscow, All-Union Institute of Scientific and Technical Information, 1970.
186. Pulyarkin, V.A. Concept of the geographical environment and the influence of the geographical environment on society. In: Nature and society (Priroda i obshchestvo), Moscow, Nauka Press, 1968.
187. Pulyarkin, V.A. Characteristics of the formation and structure of economic regions in the developing countries during the period of their colonial dependency. In: Problems of geography (Voprosy geografii), Coll. 76, Moscow, Mysl' Press, 1968.
188. Pulyarkin, V.A. Lenin's territorial analysis of the development of capitalism in Russia and the study of the geography of the developing countries, Izvestiya (Bulletin) of the Academy of Sciences of the USSR, Seriya Geografiya, 1970, No. 3.
189. Pulyarkin, V.A. and Tolokonnikova, L.A. Productive resources in the developing countries. In: Geographical collection (Geograficheskiy sbornik), All-Union Institute of Scientific and Technical Information, 1970, Issue 4.
190. Development of small and medium-sized cities of the central economic regions of the USSR (Puti razvitiya malykh i srednikh gorodov tsentral'nykh ekonomicheskikh rayonov SSSR), Moscow, Nauka Press, 1967.
191. Pchelintsev, O.S. Economic basis of the allocation of production. Methods used in the capitalist countries (Ekonomicheskoye obosnovaniye razmeshcheniya proizvodstva. Metody, primenyayemyye v kapitalisticheskikh stranakh), Moscow, Nauka Press, 1966.
192. Development and allocation of the productive resources of economic regions of the USSR (Razvitiye i razmeshcheniye proizvoditel'nykh sil ekonomicheskikh rayonov SSSR), Moscow, Nauka Press, 1967.
193. Rakitnikov, A.N. Geography of agriculture. Problems and methods of research

(Geografiya sel'skogo khozyaystva. Problemy i metody issledovaniya), Moscow, Mysl' Press, 1970.
194. Settlement in cities. Quantitative regularities (Rasseleniye v gorodakh. Kolichestvennyye zakonomernosti), Moscow, Mysl' Press, 1968.
195. Rodoman, B.B. Mathematical aspects of the formalization of regional geographical characterizations, Vestnik (Journal) of Moscow State University, Seriya Geografiya, 1967, No. 2.
196. Rodoman, B.B. Anthroposphere and complex geography. In: Nature and society (Priroda i obshchestvo), Moscow, Nauka Press, 1968.
197. Rodoman, B.B. Human activity and social-geographical regions, Vestnik (Journal) of Moscow State University, Seriya Geografiya, 1969, No. 2.
198. Rodoman, B.B. Recreational resources and the structural regionalization of recreation. In: Geographical problems in the organization of recreation and tourism (Geograficheskiye problemy organizatsii otdykha i turizma), Moscow, 1969.
199. Rozin, M.S. Geography of mineral resources in the capitalist and developing countries (Geografiya poleznykh iskopayemykh kapitalisticheskikh i razvivayushchikhsya stran), Moscow, Mysl' Press, 1966.
200. Rozin, M.S. Role of mineral resources in the formation of economic regions in the developing countries. In: Problems of geography (Voprosy geografii), Coll. 76, Moscow, Mysl' Press, 1968.
201. Saushkin, Yu.G. Soviet economic geography. In: Economic geography in the USSR (Ekonomicheskiy geografiya v SSSR), Moscow, 1965.
202. Saushkin, Yu.G. Energy production cycles, Vestnik (Journal) of Moscow State University, Seriya Geografiya, 1967, No. 4.
203. Saushkin, Yu.G. Prediction in economic geography, Vestnik (Journal) of Moscow State University, Seriya Geografiya, 1967, No. 5.
204. Saushkin, Yu.G. Mathematical methods in geography. In: Mathematical methods in geography (Matematicheskiye metody v geografii), Moscow, Moscow State University Press, 1968.
205. Saushkin, Yu.G. Geographical prediction of anthropogenic processes, Geografiya v shkole, 1968, No. 3.
206. Saushkin, Yu.G. Territorial combinations of energy production cycles, Vestnik (Journal) of Moscow State University, Seriya Geografiya, 1968, No. 1.
207. Saushkin, Yu.G. Vvedeniye v ekonomicheskuyu geografiyu. 2nd ed., supplemented and corrected. Moscow, Moscow State University Press, 1970.
208. Saushkin, Yu.G. Results and prospects for the application of mathematical methods in economic geography. In: Materialy V s"ezda Geograficheskogo obshchestva SSSR, Leningrad, 1970.
209. Saushkin, Yu.G., Smirnov, A.M. Geosystems and geostructures, Vestnik (Journal) of Moscow State University, Seriya Geografiya, 1968, No. 5.
210. Saushkin, Yu.G., Smirnov, A.M. Role of Lenin's ideas in the development of theoretical geography, Vestnik (Journal) of Moscow State University, Seriya Geografiya, 1970, No. 1.
211. Sdasyuk, G.V. Development of the concept of economic regionalization in India,

Izvestiya (Bulletin) of the Academy of Sciences of the USSR, Seriya Geografiya, 1967, No. 2.
212. Sdasyuk, G.V. Industrialization in the formation of the economic regions of India, In: Countries and peoples of the East (Strany i narody Vostoka), Moscow, 1967, Issue 5.
213. Semevskiy, B.N. Development of the economic geography of the capitalist countries in the Soviet Union, Vestnik (Journal) of Leningrad State University, 1967, No. 12, Issue 2; No. 18, Issue 3.
214. Semebskiy, B.N. Methodological bases of geography, Vestnik (Journal) of Leningrad State University, 1968, No. 24.
215. Semevskiy, B.N. Review of Nature and Society (Priroda i obshchestvo), a collection of articles, Moscow, Nauka Press, 1969. Izvestiya (Bulletin) Vsesoyuznogo Geograficheskogo Obshchestva, 1969, No. 2.
216. Semevskiy, B.N. Unified geography propaganda, Izvestiya (Bulletin) Vsesoyuznogo Geograficheskogo Obshchestva, 1969, No. 6.
217. Semevskiy, B.N. Lenin's role in the development of economic geography, Vestnik (Bulletin) of Leningrad State University, 1970, No. 6.
218. Semevskiy, B.N. Interaction of man and nature, Priroda, 1970, No. 8.
219. Semonov, P.Ye. Formation of the regional productive complexes of Kazakhstan. In: Problems of geography (Voprosy geografii), Coll. 80, Moscow, Mysl' Press, 1970.
220. Silayev, Ye.D. Productive-territorial complexes (Proizvodstvenno-territorial'nyye compleksy), Baku, Azerneshr Press, 1968.
221. Social-economic maps in complex regional atlases (Sotsial'no-ekonomicheskiye karty v kompleksnykh regional'nykh atlasakh). A collection of articles, Moscow, 1968.
222. Sochava, V.B. Practical significance of geographical research and the concept of applied geography, Doklady (Reports) Instituta geografii Sibiri i Dal'nogo Vostoka, 1965, Issue 9.
223. Stepanov, M.N. Development of the concept of the energy production cycle. In: Problems of geography (Voprosy geografii), Coll. 75, Moscow, Mysl' Press, 1968.
224. Stepanov, M.N. Socialist territorial organization of productive resources and the development of the concept of the productive-territorial complex. In: Productive-territorial complexes. Geography of the USSR (Proizvodstvenno-territorial'nyye kompleksy. Geografiya SSSR), Issue 8, Results in science (Itogi Nauki), Moscow, All-Union Institute of Scientific and Technical Information, 1970.
224a. Stepanov, M.N. Stages in material production as structural elements in the territorial-spatial organization of productive resources, Uchenyye zapiski (Scientific Notes) of Perm' State University, 1970, No. 242.
225. Tverdokhlebov, I.T. Foundations of recreational regionalization. In: Geographical problems in the organization of recreation and tourism (Geograficheskiye problemy organizatsii otdykha i turizma), Moscow, 1969.
226. Thematic cartography in the USSR (Tematicheskoye kartografirovaniye v SSSR), Leningrad, Nauka Press, 1967.
227. Territorial organization of the productive resources of the USSR. In: Problems of

geography (Voprosy geografii), Coll. 75, Moscow, Mysl' Press, 1968.
228. Territorial productive complexes. In: Problems of geography (Voprosy geografii), Coll. 80, Moscow, Mysl' Press, 1970.
229. Utkin, G.N. Theory and practice of the integral economic regionalization of the countries of Africa, as illustrated by Morocco. In: Problems of geography (Voprosy geografii), Coll. 76, Moscow, Mysl' Press, 1968.
230. Feygin, Ya.G. Lenin and the socialist allocation of productive resources (Lenin i sotsialisticheskoye razmeshcheniye proizvoditel'nykh sil), Moscow, Nauka Press, 1969.
231. Freykin, Z.G. Lenin's ideas on the economic regionalization of the USSR and regional economic geography, Geografiya v shkole, 1970, No. 2.
232. Khodzhayev, D.G. and Khorev, B.S. Concept of a unified system of settlement and the planned regulation of the growth of cities in the USSR. In: Problems of urbanization in the USSR (Problemy urbanizatsii v SSSR), Moscow, Moscow State University Press, 1971.
233. Khorev, B.S. Urban settlement in the USSR. Problems of growth and their study. Outline of settlement geography (Gorodskiye poseleniya SSSR. Problemy rosta i ikh izucheniye. Ocherki geografii rasseleniya), Moscow, Mysl' Press, 1968.
234. Khrushchev, A.T. Progress in the territorial organization of industry in the USSR. Methodology and practice. In: Problems of geography (Voprosy geografii), Coll. 75, Moscow, Mysl' Press, 1968.
235. Khrushchev, A.T. Industrial geography of the USSR (Geografiya promyshlennosti SSSR), Moscow, Mysl' Press, 1969.
236. Khrushchev, A.T. Lenin and the territorial organization of industry in the USSR, Geografiya v shkole, 1969, No. 5.
237. Khrushchev, A.T. Industrial blocks of the USSR and the principles of their typology, Vestnik (Journal) of Moscow State University, Seriya Geografiya, 1970, No. 2.
238. Chetyrkin, V.M. Problems in economic regionalization (Problemnyye voprosy ekonomicheskogo rayonirovaniya), Tashkent, 1967.
239. Sharygin, M.D. Stages in the development of territorial-productive complexes. A formulation of the problem, Uchenyye zapiski (Scientific Notes) of Perm' State University, 1970, No. 211.
240. Shkurkov, V.V. Cartographical appraisal of the natural conditions of life of the population of Northern Kazakhstan, Vestnik (Journal) of Moscow State University, Seriya Geografiya, 1967, No. 5.
241. Shrag, N.I. Industrial complexes. A theoretical outline (Promyshlennyye kompleksy. Teoreticheskiye ocherki), Moscow, Ekonomika Press, 1969.
242. Economic geography of foreign countries (Ekonomicheskaya geografiya zarubezhnykh stran), Part 1, Moscow, Prosveshcheniye Press, 1968.
243. Economic geography of the capitalist countries of Europe (Ekonomicheskaya geografiya kapitalisticheskikh stran Yevropy), Moscow, Moscow State University Press, 1966.
244. Economic geography of the Soviet Union. Part 1. The productive resources of the country as a whole (Ekonomicheskaya geografiya Sovetskogo Soyuza. Chast' 1.

Proizvoditel'nyye sily strany v tselom), Moscow, Moscow State University Press, 1967.
245. Economic problems in the allocation of the productive resources of the USSR (Ekonomicheskiye problemy razmeshcheniya proizvoditel'nykh sil SSSR), Moscow, Nauka Press, 1969.
246. Economic regionalization of the developing countries. In: Problems of geography (Voprosy geografii), Coll. 76, Moscow, Mysl' Press, 1968.
247. Encyclopedic dictionary of geographical terms (Entsiklopedicheskiy slovar' geograficheskikh terminov), Moscow, Sovyetskaya Entsiklopediya Press, 1968.
248. Epshtein, A.S. Interaction of industries in territorial complexes. In: Economic-geographical problems in the development of the territorial-productive complexes of Siberia (Ekonomiko-geograficheskiye problemy formirovaniya territorial'no-proizvodstvennykh kompleksov Sibiri), Novosibirsk-Irkutsk, 1970.
249. Anucsin, V.A. A szovjet kiro földrajz elvi alapýać, Földr. Kozl., 1966, No. 3.
250. Maergoiz (Mayergoyz), I.M. Regarding the typology in economic geography, Peterm., Geogr. Mitteilungen, 1967, No. 3.

THE APPLICATION OF MATHEMATICAL METHODS IN ECONOMIC GEOGRAPHY

Y. G. Lipetz

The use of mathematical methods in geographical research has rapidly increased in the past few years. This process did not occur in isolation but as an integral part of the development of geography as a whole in the contemporary stage of the scientific and technological revolution.

Not only a broad spectrum of new mathematical and technical means of analyzing systems but also large computers (of the BESM-6 type) and mathematical–economics methods and models played an important role in the development of theoretical investigations and in the solution of many practical problems in planning. As was noted in the report of the Central Committee of the Communist Party of the Soviet Union to the 23rd Congress of the Party, "Science has significantly enriched the theoretical arsenal of planning and has developed mathematical–economic modeling, systems analysis, and other procedures."*

Many geographical studies dealing with the formulation of practical recommendations were carried out, as a rule, in close contact with economists, while, at the same time, a number of theoretical investigations were proceeding independently and relying to a significant degree on new mathematical procedures. For various reasons — the greater complexity of the systems studied in geography, and the fact that they have not been adequately studied and structured — geography to a significant degree lagged behind a number of related economic and natural-science disciplines in the application of mathematical methods. In order to eliminate this lag geographers took a number of measures: they organized seminars and summer schools, taught courses in universities, and participated in collaborative studies with specialists from other areas. The education and re-education of workers in the field

---

*L. I. Brezhnev. "Outline Report of the Central Committee of the CPSS to the 23rd Congress of the Communist Party of the Soviet Union," Moscow, 1971, p. 83.

proceeded at a rapid rate, and at the same time theoretical, methodological, and practical research was carried out.

The first course in the application of mathematical methods was given in 1960-1961 in the Department of Geography at Moscow State University by a young scholar, now a member of the Academy of Sciences of the USSR, Aganbegyan. More specialized courses were latter offered at Odessa University, Tartu University, Kazan University, Novosibirsk University, and elsewhere. Kazan University was the first to train specialists in mathematical geography who were well acquainted with modern computer technology and programming methods.

The development of this line of research was facilitated in many respects by a seminar in new methods of economic–geographic research organized by the Economic Geography Section of the Moscow Branch of the Geographical Society of the USSR. Specialists from Moscow University, various institutes of the Academy of Sciences of the USSR, and Gosplan – the Institute of Geography of the Academy of Sciences of the USSR, the All-Union Institute of Scientific and Technological Information, the Institute of Complex Transportation Problems, the Central Institute of Mathematical Economics, the Council for the Study of Productive Resources, the Central Scientific Research Institute of Economics – and various planning organizations participated. Most of the pioneering studies in the area of application of mathematical methods performed during the period under consideration were described and discussed at this seminar, and the first lecture course on the use of mathematical methods in geography was organized (by Vasilevskiy), attracting a wide audience of graduate specialists.

An important part of the re-education of specialists working in this area was played by the three summer schools ("Mathematics in Geography") organized by the Department of Geology of Moscow State University (Saushkin, chairman of the organizing committee) in cooperation with Kazan (1966), Tartu (1967), and Lvov (1968) Universities.

Material from completed studies was presented at the First All-Union Interdepartmental Conference ("Mathematical Methods in Geography"), in which 200 geographers from 30 cities and representatives of 62 research, educational, and planning organizations took part.

Also of considerable importance was the All-Union Conference on the Application of Mathematical Economics Methods and Computer Technology in Developmental and Allocation Planning, held in 1967 in Talinin. At this conference summaries were presented on a wide range of studies in the area of application of new planning methods, as well as a scheme for a standard methodology; the latter was subsequently approved by Gosplan USSR and the Academy of Sciences of the USSR and used as the basis for many allocation models for branches of the national economy [17].

In addition to a large number of articles in scientific periodicals, this period saw the publication of the first books in this field, mainly collections of articles. The first collection of articles from the seminar on new methods was published in 1964 [5]. The second and third editions of Medvedkov's book [11] primarily dealing with the application of mathematical methods in population geography came out in 1965-1966. The papers read at the Talinin conference were published in 1967 in four volumes [10]; a collection of papers read at the Moscow conference (*Mathematical Methods in Geography*) was published in 1968 [9], as was the first collection of articles in the series *Problems of Geography, Mathematics in Economic Geography* [8].

After 1968 the introduction of mathematical methods into economic geography continued in a somewhat different form. In addition to attracting an ever growing number of investigators, studies in this area applied mathematics and computers to a wider range of problems. The initial stage of development of this field had ended: in various cities and scientific centers schools had formed that concentrated their attention on individual economic–geographic and mathematical problems. If previous research had largely been the result of the initiative of individual workers, now a natural differentiation of work in the field was taking place. The most promising lines of research became part of the agendas of a number of organizations, and a number of workers in the field who had accumulated experience in applying mathematical methods and who possessed organizational skills headed entire teams of specialists.

The teams of specialists working in the departments of geography at Moscow, Kazan, Tartu, Odessa, Novosibirsk, Irkutsk, and Riga Universities, and at the Institutes of the Academy of Sciences and Gosplan – the Institute of Geography of the Academy of Sciences of the USSR, the Institute of Economics and Organization of Industrial Production (of the Siberian Section of the Academy of Sciences of the USSR), IGSVD of the Siberian Section of the Academy of Sciences of the USSR, the Central Scientific Research Institute of Economics of Gosplan – may be included in the above-mentioned group of schools, although clearly this list is not complete in many organizations there are men who are becoming independent condensation nuclei for new teams. This process of rapid development is also evoking interest abroad. The most representative of foreign surveys on this subject is the article by Jensen and Karaska entitled "The Mathematical Thrust in Soviet Economic Geography" [22]. In evaluating the achievements of Soviet economic geography, on the basis primarily of works published from 1964 to 1967, the authors drew several correct conclusions, in particular, regarding the strong link between the use of mathematical methods and the general methodological conceptions and Marxist world view of Soviet economic geography. A more complete and valid evaluation of the basic factors influencing the introduction of mathematical methods into Soviet economic geography and the development of a number of schools and lines of research that differ from each other and from foreign (especially bourgeois) schools is now possible, however.

There are many factors and preconditions that explain the rapid development of the application of mathematical methods in Soviet economic geography.

The teachings of Marxism–Leninism facilitated the unification within the scope of economic geography of many disciplines which are designated abroad by such terms as human geography, regional science, ekistics, urban studies, etc. This unification, though incomplete (in its social aspects), and the trend in certain disciplines, population geography for example, towards greater autonomy nevertheless facilitated the development both of general conceptions and of their formalization at a higher level of abstraction, which constituted a necessary stage in the mathematization of the field.

An important methodological condition for mathematical modeling in economic geography is the scientific isolation of the basic motive forces and factors of economic development described in the works of the classic Marxist-Leninist authors and their successors. Marxists have always accorded an important place to the application of mathematics in science; Karl Marx's statements regarding the significance of mathematics in the development of science are well known.

A second important factor in this regard was the close connection between economic geography and the theory and practice of national economic planning in the USSR beginning in the first year of Soviet power with the State Commission for the Electrification of Russia plans, the five-year plans, the Ural–Kuznetsk Kombinat, and the Volga and Angara hydroelectric stations, and ending with the long-range General Scheme for the Development and Allocation of the Productive Resources of the USSR.

A third positive factor was undoubtedly the development of Soviet mathematics and computer technology on the one hand and the major successes of economic science on the other. The development of the broad field of mathematical programming (Kantorovich), together with the considerable achievements of a number of neighboring disciplines, the wide application of new methods in economic models (Nemchinov) at all levels of the national economy, the creation of major specialized academic institutes in Moscow and Novosibirsk, and the emergency of a large number of laboratories doing mathematical economic research in educational, scientific, and planning organizations, as well as factories and other organizations, created a favorable environment for the activities of economic geographers.

However, all of these external factors would not have been able to play their role if Soviet economic geography had not possessed the methodological foundations for the application of mathematical methods. The founders of Soviet economic geography, especially Baranskiy and Kolosovskiy, laid the theoretical basis for the introduction of the new methods. This applies equally to the general theory of economic planning, to a number of areas within economic geography, to the study of territorial–productive complexes, and to the investigation of the role of urban settlements. And it is now becoming ever more evident that the detailed study, description, and characterization of specifically economic–geographic objects can play an important role in the construction of models and estimates.

Studies of the modeling of economic allocation were logically the first to be undertaken. The use of mathematical programming techniques permitted the unification within the scope of a single model of the problems of development and allocation of production, which hitherto had been considered independently of one another and even regarded as separate disciplines. Models of this type were usually based on extreme-case analysis of an enterprise, group of enterprises, or branch of production for a given state of the rest of the national economy.

Thanks to improved programming methods and the introduction of new computers, the range of solvable problems broadened to cover about 70 branches of production, involving more than half of the total capital investment in the national economy.

Since an obvious basic shortcoming in the solution of this class of problem was a certain level of noncorrespondence between the conditions prevailing in a given branch of production, and over-all ideal conditions, investigators attempted to decrease the discrepancy between them by examining a large number of alternatives and introducing more restrictions. While participating in these efforts, geographers as a rule made especially significant corrections in these solutions in such a way as to better account for the interaction with external factors — the natural environment, population (settlement patterns and labor resources), as well as other elements of the national economy. In addition, economic geographers performed more or less independent studies of a scientific and practical nature. In the early 1960s Mikheyeva applied mathematical methods to the solution of problems involving allocation of agricultural production and specialization of agricultural regions [12]. Geographers working at the Central Scientific Research Institute of Economics of Gosplan USSR, which Mikheyeva headed, subsequently took an active part in the solution of many problems pertaining to the branch structure of production.

Geographers at the Novosibirsk Institute of Economics and the Siberian Division of the Academy of Sciences of the USSR, working together in the area of the modeling of territorial–productive complexes (under the direction of Bandman), published a number of in-depth modeling studies of Siberian TPC's [16, 21]. These studies followed closely the links between the TPC's and the territory in which they are located by modeling the regional planning for this territory and shifting to a system of intraregional models.

Although some success has been achieved in the solution of branch and regional problems, there remains the intriguing global problem of the creation of a model of the development and allocation of productive resources which unify all local (branch and regional) optimum conditions into a single system. One of the first proposals for a model of this type was made at Moscow State University by Budtolayev, Novikov, and Saushkin* and later modified by Zaitsev [8, 9]. The Council for the Study of Productive Resources [23] and the Institute of Economics and Organization of Industrial Production [14] are now conducting similar studies. The Central Institute of Mathematical Economics is also following several lines of research in this area. Baranov, Danilov-Danilyan, and Zavelskiy [1] are developing a system of long-range planning models in which regional models and a transport model are constructed separately. Economic geographers are actively participating in this research.

Work in this area has already developed beyond the level of pure abstraction. In the future their merits will be evaluated by a strict judge — the practical requirements of economic planning. And there is no doubt that, as the transition from theory to practice is made, the role of economic geography will increase even more. Even now it is not always possible to clearly distinguish between the purely economic and the spatial aspects of this research. A typical example of this is the work of the Kiev school of cyberneticists, in which geographers (Polonsky *et al.*) are actively participating in the construction of automated data collection and data-processing control systems.

In addition to their use in applied studies, mathematical methods are finding ever wider application in theoretical research. Central in this area are axiomatic constructions in geography and a new mathematical geography [8]; the work of Gurevich [8, 9] on geographical differentiation and homogeneity of geographical structures; and the work of Vasilevskiy on basic trends in the application of mathematical methods in economic geography [5, 7].

The research based primarily on material from population geography is relevant in this regard.† Medvedkov, beginning with a modification of the well-known models of Zipf and Stuart, subsequently arrived at a number of original results in the study of settlement patterns and the microgeography of cities. His pioneering work in the application of the concepts of entropy, symmetrical graphs, etc., in the analysis of settlement patterns received wide recognition abroad, where a significant amount of work in a similar direction is being done.

In the early 1960's Blazhko (Odessa University) began the study of urban settlements with the aid of mathematical methods. His work combined analysis of urban systems with the general principles of development of territorial–productive complexes and established the importance of activities of

---

*See their article in the Moscow State University Bulletin, *Geography*, 1964, No. 4.

---

†Although from the first application of mathematical methods and computers problems of allocation of production frequently became an area in which the efforts of economists and economic geographers were joined, problems of allocation of population, services and recreation facilities are still for the most part the sphere of economic geography, although here too the influence of economists and demographers is being felt to an increasing degree.

a city-forming nature in the growth of cities. These original investigations of the systems that cities constituted and of the degree to which their individual elements are interconnected are summarized in a monograph by Blazhko, Grigor'yev, and Zabotin [2].

In research of a similar nature conducted at Tartu University [8, 9], greater attention is devoted to questions involving the complex modeling of a territory, using interaction models; this research has focused, in particular, on the objective determination of regions in the Estonian SSR and the spheres of influence of individual centers.

At the Central Institute of Mathematical Economics interesting results in the modeling of various types of network systems, important in the study of spatial aspects of territorial and regional organization and in the study of population migrations in the USSR, have been obtained by Matlin and Nutenko [8, 9] and Tatevosov [9], respectively.

In addition to efforts to analyze and predict transport flows for planning purposes, among which those of the Gosplan USSR Institute of Complex Transportation Problems (Kazanskiy, Lasis, Gank, Gusev, et al.) were especially thorough and detailed. Vasilevskiy, Chizhov, and Shmelev have begun work on theoretical modeling of transport–geographic space.

Mathematical analysis of various types of geographical space is already being performed in economic geography. This is considerably facilitated by a significant flow of new studies in thematic cartography. Particularly noteworthy in this regard are papers by Chervyakov, Trunin, Serbenyuk, and Evteyev.* Recent papers by Serbenyuk and Matlin described their successful experiments in direct computer compilation of economic maps.

Thus, during the first stages of the mathematization process, especially in the solution of allocation problems, economic geography utilized the mathematical apparatus that had been created and tested in other sciences – economics, physics, and biology. The most widely used methods are those of matrix algebra and mathematical programming. They are used in a wide range of equilibrium and extreme-case models in the analysis and planning of the Soviet economy. Simplified linear programming methods were eventually replaced in the solution of allocation problems with complex modification of this approach and integer programming models. In order to facilitate calculations, in addition to the most modern computers, various mathematical–economic procedures are used, the most widely known of these being the decomposition method in the solution of multistage problems and the coordination of different planning stages in both the branch and regional sectors. It is these procedures that will serve as the basis for the nationwide production development and allocation systems

---

*Many of these were presented at the First Conference on New Developments in the Subject Matter, Content, and Methods of Compilation of Economic Maps," MFGO, 1968, p. 14.

which the Central Institute of Mathematical Economics, the Council for the Study of Productive Resources, and the Institute of Economics and Organization of Industrial Production are now elaborating. In addition, these tools are being widely applied in the solution of the most varied branch and local problems, including the development of methods for the economic appraisal of natural resources [4], the appraisal of major independent projects, and the organization of regional and urban planning.

The methods of probability theory and mathematical statistics have found application for the most part in analytic investigations. Correlation and regression analysis are encountered in many studies, but most often in the area of population geography. Factor analysis is used in the area of regionalization [5] and the statistical assay method in the modeling of the development of networks of populated points (Matlin). At the same time, the statistical methods of information theory, in particular, those used to determine entropy in spatial systems, were utilized in theoretical studies in geography [8, 9] and economic cartography [14].

Regression equations are a possible basis for one of the fundamental models of spatial interaction — the gravitation model. Of greater interest, however, is the possible use of the entire system of models of spatial interaction (gravity, energy, and electrical potential). Soviet economic geographers have done significant work in the modification of these models and the interpretation of their parameters.

Of even greater interest for geographers are the methods of graph theory and the geometry of numbers. The modeling of networks and flows in networks has a clearly geographical aspect and a number of original results have been obtained in this direction (Visilevskiy, Medvedkov, and Nutenko).

Soviet economic geography in general showed greatest interest in methods that permit solution of practical problems in the planning of the national economy. The arsenal of mathematical methods applicable in this area increases with the development of theoretical and analytical research, but this process is nevertheless not developing as it should. In our view at least three factors account for this. 1) We do not as yet have a manual of mathematical methods or a textbook for use in institutes and universities Chervyakov's little book [20] was published by the Far Eastern University in an edition of only 500 copies, while a pamphlet on various topics [18], published by Kazan University, came out in an edition of only 200 copies. Translations — the books by Izard [3] and Hagget [19] and the collection of articles *Models in Geography* [13] — play a certain role, but they fail to cover a number of important topics and are by nature poorly suited to a socialist society. 2) There exist no methodological centers developing and applying mathematical methods in geography such as those which exist in economics. 3) Too little attention is devoted to mathematical methods in the agendas of a number of major scientific and educational institutes.

Nevertheless, mathematical methods are being used more and more widely and appear to have good long-term prospects. Geographers are assimilating new areas of theoretical and applied mathematics (see, for example, the work on theory of reserves done by Blazhko and his coworkers, reported in the above-mentioned monograph). Geographers are attracted to research on many boundary problems. It is entirely possible that the mathematization of geographic theory will resolve the dilemma as to whether to attack allocation problems directly, by constructing a complex system of optimization models, which will deterministically account for an enormous number of indicators, factors, and parameters, or to attempt to achieve serious results in the general theory of allocation. Success in this line of inquiry is causing a shift in the understanding of mathematization as a purely methodological process, even simply as the introduction of new calculation procedures, that a number of the important theoretical conceptions advanced by Soviet economic geography are coming to be regarded from the point of view of modern mathematics (set theory, mathematical logic, and other general disciplines).

LITERATURE CITED

1. Baranov, E.F., Danilov-Kanil'yan, V.I., Zavel'skiy, M.G. Development of a system for the optimal planning of the national economy, Moscow, TsEMI of the Academy of Sciences of the USSR.
2. Blazhko, N.I., Grigor'yev, S.V. and Zabotin, Ya.I. Mathematical-geographic methods for investigating urban settlements (Matematiko-geograficheskiye metody issledovaniya gorodskikh poseleniy), Kazan State University Press, 1970.
3. Izard, U. Methods of regional analysis (Metody regional'nogo analiza), Moscow, Progress Press, 1966.
4. Initial assumptions of a methodology for the economic appraisal of natural resources (Iskhodnyye polozheniya metodiki ekonomicheskoy otsenki prirodnykh resursov), Moscow, TsEMI of the Academy of Science of the USSR, 1970.
5. Quantitative research methods in economic geography (Kolichestvennyye metody isslodovaniya v ekonomicheskoy geografii), Moscow, All-Union Institute of Scientific and Technical Information – MFGO, 1964.
6. Kossov, V.V. Interbranch balance (Mezhotraslevoy balans), Ekonomika Press, 1966.
7. Short geographical encyclopedia (Kratkaya geograficheskaya entsiklopediya), Moscow, 1966, pp. 149-159. D.L. Armand, Mathematical methods in physical geography; L.I. Vasilevskiy, Mathematical methods in economic geography.
8. Mathematics in economic geography. Problems of geography (Voprosy geografii), Coll. 77, 1968.
9. Mathematical methods in geography (Matematicheskiye metody v geografii), Abstracts of papers read at the First All-Union Interagency Conference on Mathematical Methods in Geography, Moscow, May 21-25, 1968, Moscow, Moscow State University Press, 1968.
10. Materials of the All-Union Conference on the Application of Economic-Mathematical Methods and Computer Technology in the Planning of the Development and Allocation of Production, Tallin, 1967, in four volumes. Section 1. Optimal branch planning of the development and allocation of production. Section 2. Optimal planning of the development and allocation of production in regions. Section 3. Mathematical methods in the optimal planning and allocation of production.
11. Medvedkov, Yu.V. Economic-geographical study of the regions of the capitalist world, Issue 2, Application of mathematics in economic geography, Moscow, All-Union Institute of Scientific and Technical Information, 1965, Issue 3. Analysis of settlement configurations. Moscow, All-Union Institute of Scientific and Technical Information, 1966.
12. Mikheyeva, V.S. Mathematical methods in planning the allocation of agricultural production (Matematicheskiye metody v planirovanii razmeshcheniya sel'skokhozyaystvennogo proizvodstva), Moscow, Ekonomika Press, 1966.
13. Models in geography (Modeli v geografii), Progress Press, 1971.
14. New developments in the subject matter, content, and methods of compiling economic maps (Novoye v tematike, soderzhanii i metodakh sostavleniya ekonomichiskikh kart), Moscow, MFGO, 1970.
15. New methods of economic-geographical research. Geography of industry (Novyye

metody ekonomiko-geograficheskikh issledovaniy. Geografiya promyshlennosti), Moscow, MFGO, 1967.
16. Optimal territorial-productive planning (Optimal'noye territorial'no-proizvodstvennoye planirovaniye), Novosibirsk, Nauka Press, 1969.
17. Basic assumptions of the optimization and allocation of production (Osnovnyye polozheniya optimizatsii razvitiya i razmeshcheniya proizvodstva), Moscow, TsEMI of the Academy of Sciences of the USSR, Institute of Economics of the Department of Industrial Establishments, Siberian Department of the Academy of Sciences of the USSR, Council for the Study of Productive Reserves of the State Planning Agency, 1969.
18. Manual for a course in mathematical geography (Rukovodstvo po kursu matematicheskaya geografiya), Kazan, Kazan State University, 1969–1970. Topic 1. Some problems and goals of mathematical geography; Topic 2. Mathematical methods used in economic geography; Topic 3. Economic-mathematical modelling of the national economy; Topic 4. Economic-mathematical modelling of territorial-productive complexes.
19. Khagget, P. Spatial analysis in economic geography (Prostranstvennyy analiz v ekonomicheskoy geografii), Moscow, Progress Press, 1968.
20. Chervyakov, V.A. Fundamentals of mathematical statistics in geography. A textbook (Osnovy matematicheskoy statistiki v geografii. Uchebnoye posobiye), Vladivostok, Far Eastern State University, 1966.
21. Economic-geographical problems in the formation of the territorial-productive complexes of Siberia (Ekonomiko-geograficheskiye problemy formirovaniya territorial'no-proizvodstvennykh kompleksov Sibiri, Issue 1, Novosibirsk, 1969; Issue 2, Irkutsk-Novosibirsk, 1970. Modelling the formation of territorial-productive complexes (Modelirovaniye formirovaniya territorial'no-proizvodstvennykh kompleksov), Novosibirsk, 1971.
22. Jensen, R.G. and Karaska, G.J. The mathematical thrust in Soviet economic geography, J. Regional Science, 1969, No. 1.
23. Regional Science Association Papers, **14, 18, 20, 22,** 1967–1970.